Sapere Aude

(Dare to Know)

An Executive Summary of the Trilogy Naturalistic Reason

Phillip Scribner

Publishing History

Paperback Edition 1 / May 2022
ISBN: 978-1-7379884-9-6
Imprint: natReason LLC
Washington, District of Columbia

Library of Congress Control Number: 2022906875

Sapere Aude

Is Dedicated to

All those who believe in the rational pursuit of truth.

Contents

Table of Figures

Introduction

Dare to know! That was the motto that Kant gave the Enlightenment when he defended it in 1784.[1] He was characterizing the spirit of his age a century after the cultural progress begun by Newton and Locke. He saw it as the courage to abandon the habit of holding beliefs because they are defended by authorities, such as the church, the state, and tradition. Kant urged everyone to use reason and judge for themselves what is true. He defended freedom of speech, expecting the public exchange of arguments to discover eventually the wisdom that philosophers loved.

Two and a half centuries after Kant, we know that the Enlightenment led to progress. The science inspired by Newton is now universally recognized as our most reliable source of knowledge. The liberal political order defended by Locke is the most widely shared measure of legitimate government. And the Enlightenment heralded an unprecedented improvement in the material conditions of life. After centuries of religious wars, brutal monarchies, and serfdom, the industrial revolution was undeniable progress. Indeed, statistical evidence of Enlightenment progress is indisputable.[2]

In its own terms, however, the Enlightenment failed.

It promised to settle our basic disputes. All of them! It was expected to lead to agreement not only about what to believe but also about which goals

---

[1] Immanuel Kant (1784) "Answering the Question: What is Enlightenment?" *Berlin Monthly*.

[2] For evidence of the kind that people expect, see Steven Pinker (2018) *Enlightenment Now* New York: Viking.

to choose and how to pursue them. All we had to do was give up the habit of following authority and think for ourselves.

But that has not happened, and confidence in the rational pursuit of truth has eroded. These days, no one expects basic disputes to be settled by reasoning. Science is far from providing the complete understanding of the natural world it promised. Physics offers theories about the nature of what exists that we cannot even understand. Biology cannot explain the nature of life, much less its origin. The regularities described by laws of physics cannot explain the more specialized regularities studied by other branches of science. It is so baffling how consciousness, the phenomenal aspect of experience, is part of the natural world, that neural scientists as well as philosophers now call it the *hard problem of mind.*

What is worse, even facts about current affairs are at issue. Political disagreements are dividing people into opposing tribes. They have different beliefs, and instead of arguing about what is true, they tend to vilify those who disagree with them. Some try to silence opposing views as hate speech—even using force, if necessary. Mass media routinely censor what they call disinformation. And it is only getting worse.

This decline in our culture has not gone unnoticed. In the shrinking number of people who are familiar with it, the Enlightenment arouses nostalgia. With a lingering faith in the rational pursuit of truth, they hold out hope of resolving political disputes peacefully. But in universities, mass media, and popular culture, *disillusionment* is replacing confidence in reasoning. As evidence accumulates that disputes about basic issues cannot be settled, it is becoming conventional wisdom that values are just projections of feelings onto an indifferent natural world. Liberal political principles established by the Enlightenment and built into the United States Constitution are commonly dismissed as a historical curiosity. Their validity is defended by Christians. But the existence of a transcendent creator of the natural world is a matter of faith, and the assumption that the soul is immortal leads them to make opposition to abortion their test of belief in a real difference between good and evil. The beauty of fine art was once seen as aspiring to perfection, but it is no longer expected to be beautiful at all. Instead of considering their validity, media treat arguments as *talking points* and *conversations* that may or may not alter beliefs. Journalists prefer doing good to reporting facts. Mass media use a narrative to explain what happens, and what does not fit in is ignored. Enlightenment is becoming *endarkenment*. Power is replacing rationality. Things are falling apart.

There is an antidote to this ominous decline of culture. It is a revolution in culture. And it can be foreseen. There is a scientific discovery that would solve all the problems of physics, unify physics, and reveal a kind of efficient cause not recognized by physics that would solve all the problems of other branches of science. This discovery in physics would also enable science to solve the hard problem of mind. By showing how consciousness is a pseudo-efficient cause that no one recognizes, the unification of science would lead to an explanation of Western philosophy that unifies science and philosophy. In short, we may be on the brink of a scientific revolution that transforms our culture. It would be a scientific explanation of everything in the world, and since everyone could accept it, it would resolve all the disagreements about basic issues that divide us. That would push the rising tide of endarkenment back into the sea of passionate incoherence. The center would hold. And science would turn out to be the wisdom to which philosophy has long aspired. It would be knowledge of the True, the Good, and the Beautiful.

This sounds too good to be true. But there is an argument that predicts it. That is what the following pages describe. In the rising tide of endarkenment, however, it takes courage to consider such an argument. That is why I use, *sapere aude*, the motto that Kant gave to the Enlightenment, to introduce this prediction. The Enlightenment was the beginning of science, and it was widely expected to explain everything in the natural world. But it takes more courage to consider an argument in defense of such a prediction these days than it did in in Kant's day. The aftermath of the Enlightenment seems to have shown that trust in rationality was misplaced. Indeed, an argument predicting that we are on the brink of a scientific discovery that resolves all our basic disagreements is so implausible that it may not seem worth the effort. Surely, it can only come from someone without enough education to know any better.

However, *Naturalistic Reason* predicts such a cultural revolution. It is a trilogy, and this short book contains an executive summary of the argument in the three volumes. They predict a *Second Enlightenment*, and I use the motto, *dare to know*, as the title because it challenges you to have the courage to take this argument seriously. Those who hold out hope for the rational pursuit of truth know that it would be wrong to dismiss this argument just because its conclusion is unlikely, and I believe that if you read *Sapere Aude*, you will want to learn more about the reasons for expecting it. *Part Two* of this book summarizes the arguments of all three volumes. But it is so hard to believe what they predict that I begin by showing how a Second Enlightenment is more likely than it seems. *Part One* shows how a single discovery in physics could enable science to explain everything found in the study of nature. That would complete the *18th-Century Enlightenment,* and *Part Two* shows how it could be the foundation for a *Second Enlightenment,* which would give us the wisdom to which philosophy has always aspired.

Dare to know!

Part One

Plausibility of the Prediction

From history we know that the First Enlightenment spawned science, liberal political institutions, and a wealth-creating economic system, all of which are still spreading around the globe. It is proof that it is not necessarily foolish to expect the rational pursuit of truth to cause progress. Indeed, some still expect it to lead to an explanation of everything—eventually. All I am asking is that you have an open mind. Instead of dismissing what I have to say out of hand, follow my argument until you come to a point where you can show that its premises or conclusions are contradicted by what perception tells us about the natural world. Or you can show that a formal fallacy makes the argument invalid. If you have any trust in rationality, you know that that is what you ought to do. If we come to a point where you know—or believe you know—that it is not reasonable to take another step, you can walk away in good conscience. But if you read *Sapere Aude*, I believe that you will find yourself wanting to consider it further and see how arguments could settle all the basic disputes that now seem unresolvable.

The scientific revolution predicted by this trilogy can be divided into two moments. The first describes how the discovery of a scientific explanation of all the regularities found in the study of nature could accomplish the original goal of science. The *Completion of the First Enlightenment* lays the foundation for the second moment, *The Second Enlightenment*, which shows how science could become the perfect knowledge to which philosophy has always aspired.

The 18th-Century Enlightenment was expected to lead to a complete explanation of what was found in the natural world. The success of Newtonian physics was well known, and Kant predicted further success when he gave it the motto, *sapere aude*. But ironically, his argument also helped cause a mistake that led to disillusionment with Reason. In physics, it led to intractable puzzles, and in other fields based on physics, scientific discoveries about Reason cast doubt on the power of science to explain everything found in the study of nature. However, if the initial success of science and its eventual disillusionment with Reason were *both* caused by a *single* mistake, a *single* empirical discovery could complete the First Enlightenment by correcting that mistake. That is the gist of the argument in the first two volumes of the trilogy, *Naturalistic Reason*. It predicts a discovery in physics that would do just that.

In a way, however, the completion of the First Enlightenment would be a Pyrrhic victory for science. The discovery of an explanation of everything found in the study of nature would itself be something found to happen in the natural world, and it would not have been explained. Such a discovery is possible only in the kind of *science* that history reveals came from Western civilization. Discoveries from other civilizations are not comparable. And the mistake that kept science from explaining everything will point to the cause of science in Western civilization. It came from Enlightenment confidence in Reason, and as Kant knew, its confidence came from *metaphysics*, the kind of argument that philosophers hoped would give them wisdom. Though Kant was a skeptic about metaphysics, he explained Reason in a way that defended its power to discover knowledge *a priori*, and that was the mistake that caused both the initial success of science and its eventual failure. Thus, if metaphysics is what enabled Western civilization to contribute science to humanity, it is possible that correcting a false belief derived from metaphysics would put us on a path to the perfect kind of knowledge to which philosophy has aspired ever since ancient Greece. When science discovers its own roots in metaphysics, it will recognize itself as Reason, the way of knowing in which metaphysicians believed. But even the pre-Socratics of ancient Greece aspired to perfect knowledge, and since scientists are naturalists like them, they will call it *naturalistic* Reason. That is the *Second Enlightenment*.

Chapter 1

Completing the First Enlightenment

Completing the First Enlightenment depends on correcting a mistake that made science successful at first but kept it from realizing its initial ambition until now. This mistake came from a confidence in Reason that had matured during centuries of exchanging arguments about metaphysics in the West, and Kant encouraged physics to accept it. It was a mistake about knowledge of mathematics.

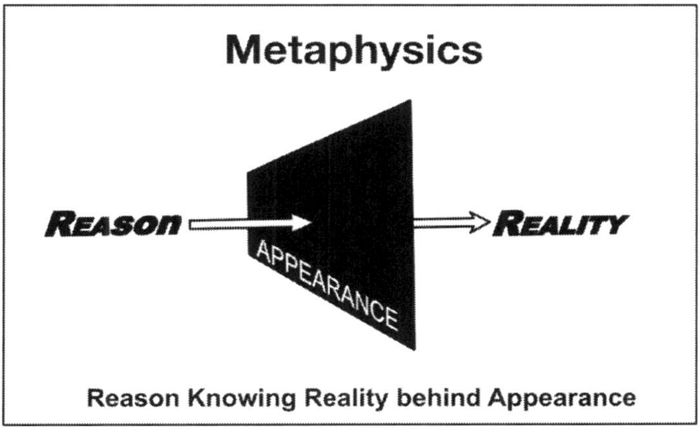

FIGURE 1 - METAPHYSICS

The basic assumption of metaphysics is that beings like us have a cognitive power, called *Reason*, that enables us to know *Reality* behind *Appearance*.[3] (See Figure 1.) This is a perfect kind of knowledge because Reason knows the nature of Reality so completely that it reveals everything that holds necessarily in Appearance, including not only the connection between causes and their effects, but also the difference

---

[3] The core terms are capitalized to distinguish metaphysical arguments from other kinds of arguments. Metaphysics has an essential nature, as science will discover.

between good and evil, not to mention what makes beautiful things beautiful. Metaphysicians used *Appearance* to refer to what seems obvious to us, for example, what we perceive in the world of objects in space where we find our bodies. But in many ways, what perception finds in that world is puzzling, and Reason is a different way of knowing about it that was expected to solve all those puzzles. Metaphysicians believe that Reason is a cognitive power that is able to know the True, the Good, and the Beautiful in a way that transcends Appearance.

Kant was not a metaphysician. Indeed, Kant rejected metaphysics. But his argument was about metaphysics, since that is what interested philosophers in his day. Reason was supposed to know Reality behind Appearance, and since Kant denied that Reason can know the nature of Reality, he called Reality the *thing-in-itself.* His *Critique of Pure Reason* defended a *constructivist* theory about the nature of Reason that implied that it could provide the perfect knowledge that philosophy had promised ever since ancient Greece without having to know the nature of Reality. He believed that Reason would enable science to succeed in explaining every kind of event in the world. His defense of the Enlightenment was practical as well as theoretical. His *Critique of Practical Reason* justified political institutions in which everyone had the freedom to which Rational beings are entitled because they are autonomous and can be trusted to be moral. But science had unprecedented success in explaining what is found in the natural world, and though Kant's theory about Reason helped cause its success, it also, ironically, helped cause the eventual disillusionment with Reason.

Science did not accept Kant's theory about the nature of Reason. Indeed, it began by rejecting a basic assumption of philosophy. Nevertheless, science did build another basic tenet of Kant's theory of Reason into its method, and it too was false. But it helped make physics successful for centuries, and after this progress, it helped cause disillusionment with Reason in science and eventually confronted physics itself with intractable puzzles. That is the kind of mistake whose correction could solve all the puzzles of physics and science and realize the initial goal of science by explaining completely what had been found in the study of nature.

What distinguished science from philosophy was its rejection of an assumption about knowledge. Like other philosophers, Kant assumed that knowledge depends on faculties of intuition, and in particular that knowledge of the natural world depends on a *faculty of perceptual intuition.* Knowers were supposed to know about the natural world because of the immediate presence of configurations of sensory qualia, or *raw feels,* such as colors and sounds. Those configurations are part of the *phenomenal character* of experience. It is now commonly called *consciousness*. But in the modern era, configurations of qualia were called *ideas*. Perceptual ideas were an Appearance behind which Reason was supposed to know Reality, and at the beginning of the modern era, metaphysics began with Descartes trying to prove the existence of the natural world behind the perceptual phenomenal appearance in order to provide a foundation of certainty for science. But science began by rejecting the assumption that knowledge of the natural world depends on perceptual *ideas*.

Science is different from philosophy because it accepts *naturalism*, defined here as the belief that *what exists is the natural world*. Naturalists assume that the world in which they find their animal bodies is made up of objects in space that exist independently of one another. This is a kind of world whose nature they understand well enough to move around and interact with other objects because, as mammals, they have a faculty of imagination built into their brain that enables them to think about the relations of objects in space and how they change as a result of moving relative to one another. I call it the faculty of *naturalistic imagination*. Input from sensory organs is registered in it, and it provides enough understanding of the natural world for mammals to survive. When mammals evolve the use of language, perception enables them to use the empirical method to learn more. They go out and look, and given what they understand about causes and their effects, they learn more about the natural world by inferring what is true as the best explanation of what they find in it.

The assumption in metaphysics and philosophy was that *all* knowledge depends on intuition, not only perception of the natural world but also reflection and everything that can be known by reflection, including Reason. I call this assumption *intuitionism*, and an intuitionistic assumption about knowledge of mathematics that science did not reject is what led to scientific progress.

At the beginning of the modern era, naturalistic defectors from intuitionistic philosophy were not called *scientists*. They were called *natural philosophers*, and they accepted what they called *mechanical philosophy*, the belief that everything in the world could be explained as a machine. Their naturalism had the same foundation as the pre-Socratic philosophers in ancient Greece, before Plato and the first metaphysical argument. In both eras, naturalists assumed that everything in the natural world is constituted by substances. *Substances* are basic causes because they are self-subsistent entities that endure through time existing in definite ways (including powers that enable them to interact with one another), and they explain what is found by constituting it. If we call substances *ontological causes,* natural philosophers and pre-Socratics are both *ontological naturalists.* But in the modern era, natural philosophers became scientists. The discovery of a new way of using the empirical method that was astonishingly successful is what turned ontological naturalists into scientists, and it compromised their ontological naturalism.

This method was used in *physics,* and in the wake of the Newtonian revolution, it was the model for other branches of science. This new way of using the empirical method can already be seen in Copernicus' defense of his heliocentric explanation of the solar system, a new geometrical explanation. It is more evident in Kepler's discovery of the *laws* of planetary motion, in which orbits were conical sections. But it was not fully explicit until Galileo used it to defend his law of falling bodies. The new method was the use of mathematics to formulate quantitatively precise laws of nature and to confirm them by careful measurements of their predictions. Galileo used a ball rolling down an inclined plane to measure the rate of change in its velocity and show conclusively that its acceleration is constant as time passes. Newton invented a new form of mathematics, called the *calculus,* to formulate his laws of motion and law of gravitation. That enabled him to predict accurately an enormously wide range of measurements, and the confirmation of Newton's laws left no room for doubt about the existence of laws of physics.

Physics had discovered a mysterious tool. Its laws described simple basic regularities about change in the natural world, and from the beginning, physicists assumed, in practice, that *mathematically formulated laws of physics are the deepest possible empirical knowledge about the natural world.* Call this method *empirical lawism.* It depends on mathematics because laws of physics use mathematics as a language for describing basic regularities about change. When predictions of physical laws are falsified, the mathematics used to formulate them is never blamed. The fault always lies with the laws, that is, in their failure to correspond to the regularities in the natural world that they purportedly describe. Their failure is known by perception, and since physicists take it for granted that mathematics is known independently of perception, they accept this distinction between mathematical theorems and laws of nature. That is what Galileo assumed when he described mathematics as the *language of nature,* and it was reinforced by Kant's argument that mathematics is known *a priori,* that is, prior to what is given in perceptual experience. But after giving physics astonishing success, it could eventually cause disillusionment with Reason in physics, if that belief were false. And it could be false. If mathematics was just a tool that depended on intuitionism, reliance on it could be a crutch that crippled ontological naturalism.

Kant's *Critique of Pure Reason* was the theory that, instead of knowing Reality behind Appearance, Reason helps construct Appearance. That is why Reason cannot know the thing-in-itself. Kant called our experience of the world the *phenomenal world,* by which he meant both the natural world and the reflective subject who perceives it, and since Reason helps construct the phenomenal world, certain aspects of it are necessary and universal. This theory about Reason is a form of intuitionism because it assumes that all knowledge depends on faculties of intuition. According to Kant, what is necessary and universal about the world of objects in space includes laws of physics as well as mathematical theorems. But mathematics is even less vulnerable to falsification because it is a more basic aspect of the phenomenal world that Reason helps construct. Kant explained space and time as forms of intuition, and he called the qualia given in those forms *matter*. He used Reason's imposition of a structure on such matter to explain why the regularities about change described by laws of physics are necessary and universal. But he explained the necessity and universality of mathematics as the product of Reason acting on the faculties of intuition by themselves, that is, apart from the (qualia) matter given in them. Geometry was the product of Reason acting on the faculty of intuition called *space*, while arithmetic was the product of Reason acting on the form of intuition called *time*.

Naturalists cannot accept Kant's theory about the nature of Reason because they assume that the natural world is Reality. Since naturalists know Reality, it is not a Kantian thing-in-itself. Indeed, Reality is something they understand in naturalistic imagination. Nor can the natural world be something presented in a faculty of perceptual intuition, like Appearance, which could be partly constructed by Reason. But since Kant's theory implies that mathematics is known independently of perception, it is more basic than the truth of laws of physics, and since that is an assumption on which the new method of physics was based, Kant's way of defending the Enlightenment, ironically, helped cause a mistake that eventually led to despair about science realizing its initial goal of explaining everything found in the natural world. Empirical lawism is a form of ontological naturalism that depends on intuitionism.

Though empirical lawism made science different from the ontological naturalism of natural (mechanical) philosophers and the pre-Socratics, science was expected to accomplish the same goal. Assuming that the natural world is constituted by substances, physicists used mathematics to describe regularities about change, and their laws of nature were expected to explain everything found in the natural world. Physics did make progress in this direction during the 18th and 19th centuries. But since the empirical method addicted them to intuitionism, this expectation had to be abandoned after the quantum and Einsteinian revolutions in the early 20th-century. The puzzles posed were so intractable that modern physicists have nearly all but given up hope of understanding in naturalistic imagination what corresponds to their mathematically formulated laws. Moreover, physics causes intractable puzzles in branches of science studying less-basic regularities. No one expects the laws of physics to explain those regularities completely. Science based on physics has lost the Enlightenment confidence about explaining everything that holds necessarily in the natural world, and its disillusionment with Reason is evident in the desperation of theoretical physicists. Some defend the *many-worlds interpretation* of quantum mechanics in which every quantum measurement divides the universe into two or more entire universes, while those who try to combine it with gravitational physics take seriously *multiverse cosmology*, in which our universe was just one of many (10^{500} or more) that began with a Big Bang. Though physicists are naturalists, their theories about what exists in the natural world are more bizarre than the ontologies, defended by intuitionistic metaphysicians, that discredited metaphysics itself!

Mathematics was both a blessing and a curse in physics. But physicists still do not realize it. Its efficacy is a mystery to physicists because they have no way to explain why mathematics is true of the natural world. As naturalists, they cannot accept what pure mathematicians say about its truth. Most mathematicians are Platonists who believe that mathematics describes a realm of entities outside space and time, like Plato's Forms, and that does not explain why mathematics holds in the natural world. Nor can physicists accept what Galileo and Newton believed about the truth of mathematics. As theists, they believed that God was a Rational being who existed outside space and time, and assuming that God used mathematics as his language for willing the natural world into existence, Galileo and Newton saw the discovery of laws of physics as a way of seeing into God's mind. But since naturalists reject supernaturalism, they cannot assume that finite mathematicians were created in God's image. What should puzzle physicists, therefore, is not only the nature of the world to which the laws of physics correspond, but also why mathematics is so *unreasonably effective* in discovering regularities that hold in it. And some confess to being puzzled by it. Famously, the Nobel-Prize winning physicist, Eugene Wigner (1960), called attention to the "unreasonable effectiveness of mathematics" in discovering laws of physics. He concluded that the "miracle of the appropriateness of the language of mathematics for the formulation of the laws of physics is a wonderful gift which we neither understand nor deserve."

Centuries of remarkable discoveries in physics have given us a level of technological power over nature that leaves no room for doubt about physics being onto something basic about Reality. It would not be all that surprising, therefore, if all the intractable puzzles about what the mathematically formulated laws of physics describe were solved by an empirical discovery that explained the "unreasonable effectiveness of mathematics" in discovering them.

Indeed, that is what *Volume I: Unification of Physics* predicts. The truth of mathematics will be explained by a discovery about space that shows what mathematics corresponds to in the natural world. It is the discovery that space is a substance that interacts with matter, and it explains the "appropriateness of mathematics" in physics because it implies that *all* the regularities generated by interactions of space and matter are quantitatively precise. In a nutshell, space has an intrinsic geometrical structure, while matter has an intrinsic quantity, and with species of bits of matter defined by the geometrical structures of their coincidence with parts of space, the powers by which space and matter interact are expressed by regular changes in the geometrical structure of their coincidence as time passes. All the basic regularities generated by interactions of space and matter are quantitative, and since they constitute all the change in the natural world, all regularities are basically quantitative. This holds regardless of the specific powers by which space and matter interact because it is a consequence of their essential natures, so, it explains the "unreasonable effectiveness" of mathematics in discovering laws of nature in any natural world constituted by space and matter. If that holds of our natural world, regularities described by laws of physics express powers of space and matter, and the puzzles of physics can be solved by a two-level inference to the best explanation. The first level infers that space and matter constitute the natural world as the best explanation of what corresponds to mathematics, and the second level infers specific powers of space and matter as the best explanation of the regularities described by laws of physics. Such powers are described in *Volume I,* and if they are on the right track, the puzzles of modern physics will be solved because naturalistic imagination will enable us to understand how interactions of space and matter generate all the regularities that physical laws describe.

To infer *spatio-materialism* as the best ontological explanation of what is found in the natural world is to make an *empirical* discovery. Like science, this inference accepts ontological naturalism, and like science, ontology uses the empirical method. But instead of assuming that laws of nature are basic, it assumes that the natural world is constituted by substances and infers their natures as the best explanation of what is found in the natural world, including not only the laws of physics but also the mathematical language used to formulate them. This discovery about the nature of space depends, therefore, on the method of *empirical ontology*, rather than *empirical lawism*. An ontological theory is inherently more basic than a physical theory because an explanation by *what exists* is more basic than an explanation of *what happens* to what exists. Since physics is the basic science, the reduction of laws of physics to interactions of space and matter will base all of science on causes of existence. Hence, the prediction is that physics will give up the crutch of intuitionism. That will turn physical science into *ontological science*.

The trilogy, *Naturalistic Reason,* predicts that this discovery about space will enable more specialized branches of science to explain the regularities that they study. Laws of nature describe regularities about change generated by interactions of substances, and since mathematically formulated laws of physics describe regularities generated by interactions of space and matter, an ontological explanation of the truth of mathematics imposes an additional constraint on what corresponds to laws of physics in the natural world. Beneath the probabilistic character of laws of quantum physics, there are deterministic ontological mechanisms that solve those quantum puzzles. But they reveal powers by which interactions of space and matter generate regularities not described by those laws. As the first volume shows, those powers are also expressed as regularities that entail a kind of efficient cause not recognized by physics, called *geometrical causes.* Geometrical causes constrain what happens as particles exert physical forces, and this combination of geometrical and physical causes is sufficient to explain all the regularities found in the study of nature. As *Volume II: Unification of Science* predicts, all the specialized branches of science will be reduced to ontology. What is ontologically necessary in the natural world will turn out to include a regularity about evolution that explains the origin of life, reveals that life has an essential nature that explains the nature of the good, and shows that a series of stages of evolution leads inevitably to the existence of language-using mammals like us, who are r*eflective subjects* because we can represent the causes of our beliefs and behavior as part of the process of causing them. But, perhaps, the most surprising implication of this unification of science is that the societies of which reflective subjects are members are a distinct form of life on a higher level of geometrical organization than multicellular animals. I call them *spiritual organisms* because they have no body but all the bodies of the reflective subjects who use language to coordinate their behavior, and that is a naturalistic explanation of our spiritual nature.

If the centuries of success in physics and the intractable puzzles that it encountered in the 20th century are both explained by an empirical discovery that implies that mathematical truth depends on correspondence to the natural world, it will not be surprising that the discovery about space being a substance that interacts with matter accomplishes the original goal of science. Nor is it implausible that this discovery will solve the puzzles of physics and reveal an unrecognized kind of efficient cause whose way of working together with physical causes unifies all the other branches of science. And if, as promised, a science based on ontology also explains the nature of the good, the unification of science could also justify offspring of the *18th-Century Enlightenment* that are based on respect for the moral autonomy of Rational beings: liberal political institutions and a capitalist economic institution. But the completion of the 18th-Century Enlightenment is the *First Enlightenment*, and my point in anticipating that conclusion here is to show how the third volume of *Naturalistic Reason* can lead to a further discovery. Those who accept the empirical method will see how a science freed of addiction to intuitionism would lead to a *Second Enlightenment* that cuts the Gordian knot of philosophical problems and turns science into naturalistic Reason.

Chapter 2

The Second Enlightenment

The prediction is that an empirical discovery that corrects a single basic assumption of physics will complete the First Enlightenment. The reduction of physics to ontology will explain all the regularities that science has found in its study of nature, including the evolution of reflective subjects like us. But this will be accomplished by a science based on ontology, and since its existence will not have been explained, there will be something that still calls for an explanation. That is the existence of ontological science itself. I call the discovery of its origin the *Second Enlightenment*.

Since ontological science comes from reducing physics to spatio-materialism, tracing the origin of ontological science will require ontological scientists to explain the origin of physical science. Science began during the 18th-Century Enlightenment with the astonishing success of physics in discovering mathematically formulated laws of nature, and as we have seen, its success depended on physics accepting a false belief about the truth of mathematics. Empirical lawism assumes that mathematics is known independently of perception, and that assumption came from philosophy. It was reinforced by Kant's constructivist theory about the nature of Reason. He held that Reason could give us the perfect knowledge sought by philosophy without knowing Reality behind Appearance. But his argument explained the truth of mathematics as an effect of Reason acting on forms of intuition called *space* and *time*, and that is false. Its truth depends on correspondence to regularities generated by interactions of space and matter that depend only on their essential natures. Thus, what perception finds in the natural world can falsify mathematics as well as laws of nature. And the puzzle about how a false philosophical belief about the nature of mathematical truth could make physics so successful will be solved when ontologists trace the origin of ontological science to metaphysics.

Though Kant held that metaphysics was bound to fail, his constructivist theory of Reason was meant to show that, nevertheless, Reason could provide the wisdom it promised, and he shared with metaphysicians and all other philosophers the assumption that knowledge depends on faculties of intuition. *Intuitionism* was the foundation of metaphysical arguments. Appearance as well as the faculty of Reason that knows Reality behind Appearance were assumed to be objects presented by faculties of intuition. But physicists were naturalists, and since they assumed that they were animals who perceive a world of objects in space that exist independently of one another, they rejected intuitionism. Assuming that knowledge of the natural world depends on sensory organs, rather than a faculty of intuition, they used the empirical method to learn more about it. Thus, when physicists defected from philosophy, they rejected one mistaken assumption of intuitionistic metaphysics. But as empirical lawists, they kept another mistaken assumption. They assumed that mathematics was known by Reason.

The history of science shows that mathematics came from intuitionism, as described in the last chapter. But that does not explain why metaphysicians and other philosophers accepted intuitionism, much less how *a priori* mathematics could be a valuable tool. This is a mystery about Western civilization because no other great civilizations on Earth discovered anything like physics, and *Volume III: Unification of Science and Philosophy* predicts that ontological scientists will discover that intuitionism depends on consciousness.

Since ontological scientists are conscious reflective subjects, they will be able to follow metaphysical arguments, and they will recognize that objects of intuition are parts of consciousness. *Consciousness* is used here as a term that refers to the phenomenal character of experience, or to use a current term of art, consciousness is *what it is like* to be a being like us.[4] Consciousness was ignored by physical scientists studying nature after the Enlightenment. But it is found in the natural world, and spatio-materialism will enable ontological scientists to explain how consciousness is part of the natural world. Their explanation will reveal what is true and what is false in metaphysical arguments, and much to their surprise, they will find that by explaining it, they have turned ontological science into Reason. They will find themselves knowing Reality behind Appearance. But since they are naturalists using the empirical method, they will call it *naturalistic Reason* to contrast it with *intuitionistic Reason*, the kind of theory that was taken for

---

[4] This way of describing consciousness was introduced by Thomas Nagel (1974) "What is it Like to be a Bat?" *The Philosophical Review*, Vol. 83, No. 4., p. 435-450.

granted in Western philosophy. Nevertheless, that will be the *Second Enlightenment* because Reason will still be the perfect kind of knowledge to which philosophers—*lovers of wisdom*—have aspired ever since ancient Greece.

Spatio-materialism will enable ontological scientists to explain how consciousness is part of the natural world if they assume that matter has a phenomenal way of existing in itself. They might convey what they mean by saying that *it is like something* to be a bit of matter. Such a theory is called *panpsychism* because it must hold of every bit of matter in the natural world. Since bits of matter have more or less complex ways of coinciding with parts of space, they can have more or less complex phenomenal ways of existing in themselves, and since configurations of sensory qualia are complex phenomenal properties, consciousness must be the property of a complex bit of matter. In perception, such phenomenal properties are immediately present, and if this is how consciousness is part of the natural world, their immediate presence is not an efficient cause. What happens in the natural world depends on efficient causes, and though phenomenal properties help constitute what exists in the natural world as it happens, their immediate presence does not make any difference in *what* happens. This kind of theory about phenomenal properties is called *epiphenomenalism*. The phenomenal intrinsic property of a bit of matter as something different from the bit of matter that has that property is causally impotent. But this causal impotence does not mean that phenomenal properties are irrelevant in explaining what happens because there is an illusion inherent in consciousness. That illusion can cause errors.

Ontological scientists will explain consciousness as a complex phenomenal intrinsic property of a bit of matter that helps constitute the *mammalian brain*. From their explanation of stages of evolution, they will know that mammals evolve because the structure of the mammalian brain serves as a faculty of imagination. Its basic structure is a geometrical cause that registers sensory input in a way that can be recalled as representations of the consequences of possible ways of behaving, and since it enables mammals to perceive themselves as animal bodies located in space along with other material bodies that exist independently of them, I call them *subjective animals*. There is a bit of matter helping constitute the mammalian brain whose way of coinciding with space has a structure that registers the activity of all the neurons in its faculty of imagination, and since it is the only bit of matter that can explain the complex phenomenal intrinsic property that is immediately present to mammals when mammals perceive the natural world, ontologists will conclude that its phenomenal way of existing in itself is what makes mammals conscious. Since this phenomenal property is isomorphic with the activity of the faculty of imagination, it naturally appears to mammals that they are aware of representations of the natural world constructed from sensory input in that faculty *because* they are immediately present, like objects of intuition. I call this the *illusion of intuitionism.*

Reflective subjects are mammals whose use of language enables them to represent their psychological states. But they are still misled by the intuitionist illusion. Besides phenomenal properties that correspond to the natural world in perception, reflective subjects can describe phenomenal properties that correspond to their psychological states, such as what they imagine, remember, dream, and feel about the objects represented in them, and the immediate presence of phenomenal properties that inhere in psychological states also seems to be what causes their knowledge of psychological states. In other words, reflective subjects naturally assume that they have both a faculty of perceptual intuition and a faculty of reflective intuition, and since their objects include both the natural world and the subject who perceives it, they explain what Kant called the *phenomenal world*.

In particular, this explanation of intuitionism as an illusion will enable ontological scientists to explain why Kant believed that mathematics is known *a priori*. As we have seen, Kant explained Reason as a cognitive power that helps construct the phenomenal world, and it acts not only on the "matter" (or qualia) given in forms of intuition, but also on the forms of intuition themselves. Thus, Kant explained geometry as a product of Reason acting on the *form of intuition* that he called *space* independently of the "matter" given in it, and he explained arithmetic as Reason acting on another *form of intuition* that he called *time* independently of the "matter" given in it. Other aspects of "matter," such as regularities described by laws of physics, were supposed to depend somehow on things in themselves as well as Reason. But mathematics does not depend on things in themselves at all, and since geometry and arithmetic are a sufficient foundation for mathematicians to derive all the theorems used in physics, Kant's explanation of how mathematics was known *a priori* justified physicists in thinking of mathematics as a language for formulating laws of nature. Mathematics could not be falsified by what perception finds in the natural world because its truth did not depend on things in themselves, that is, on correspondence to what naturalists consider Reality: the natural world. This encouraged physicists to believe that mathematics was just a language in which they could describe the most basic regularities about change found in the natural world.

Kant allowed physicists to believe that mathematics was immune from falsification by perception. But as we have seen, the use of mathematics began earlier, with Copernicus, Kepler, and Galileo. Indeed, the calculus that enabled Newton to discover his law of gravitation and three laws of motion depended on the use of coordinate systems that Newton learned as a young man from his study of Descartes' work on geometry.[5] Descartes was the metaphysician who began the modern era, and he explained Reason as a faculty of intuition on the model of the faculty of perceptual intuition. Ontological scientists will know that objects of perceptual intuition are phenomenal properties that inhere in representations of the natural world constructed in the faculty of naturalistic imagination, and likewise, they will explain mathematical objects of rational intuition as phenomenal properties that inhere the psychological states of language-using mammals using their faculties of naturalistic imagination to make arguments about the structure of space and the numbers used to count objects in space. But in both cases, Descartes was taken in by the illusion of intuitionism. Objects in the faculty of perceptual intuition were phenomenal properties of representations of contingent states of affairs in the natural world, but since objects in the faculty of rational intuition were phenomenal properties of inferences that had to hold in the faculty of naturalistic imagination, they seemed to be self-evident. To contrast them with ideas that could be doubted, Descartes described these mathematical ideas as clear and distinct—*clear* because every detail was immediately present, and *distinct* because their self-evidence did not depend on anything but what was immediately present. Since they were indubitable, they could serve as a foundation on which to base theorems of geometry and arithmetic. Descartes was not puzzled by the correspondence of the clear and distinct mathematical ideas to the natural world because he believed that Reason came from God creating finite beings in his own image, and he agreed with Galileo about God using mathematics to will the natural world into existence.

Though Descartes and Kant could both explain why mathematics was not falsifiable by perception, Descartes offered a better explanation of how false beliefs about mathematics made physics successful. Since Kant's

---

[5] It is contained in the final essay of Descartes' *Discourse on Method*, published in 1637.

constructivist theory implied that Reality was an unknowable thing-in-itself, it could not explain the unreasonable effectiveness of mathematics in discovering laws of physics in a world where its truth depends on correspondence to Reality. But in Descartes' theory of rational intuition, the clear and distinct ideas of mathematics correspond to the basic structure of the natural world because they are the phenomenal appearances of naturalistic imagination when that faculty is used to think about relations of objects in space and how they change when objects move. Indeed, ontological scientists will recognize that Descartes' use of numbers to represent distances in three independent coordinates corresponds to the structure of space. Knowing that the truth of mathematics depends on its correspondence to any world constituted by interactions of space and matter, they will understand why trust in Cartesian rational intuition made physicists so successful in using perception to confirm mathematically formulated laws of nature. Indeed, ontological scientists will explain mathematics as a kind of knowledge that depends on perception even though it cannot be falsified in a spatio-material world. They will know that the mammalian brain acquires its faculty of naturalistic imagination because as children play, a process of neurological development internalizes the structure by which space gives objects relations to one another and how they change as they move relative to one another. In short, ontological scientists will discover how a false belief about how mathematics is known could make mathematics so *unreasonably effective* as a language for formulating laws of physics.

This way of tracing the ancestry of ontological science through physical science to Descartes will not be a complete explanation of its existence because it will not explain why he defended a theory of Reason. Though the faculty of rational intuition enabled him to show how Reason knows Reality behind Appearance, that does not explain why he made a metaphysical argument. Since the first metaphysical argument was made by Plato in ancient Greece, modern metaphysics was a product of nearly two millennia of arguing about metaphysics. Western philosophy was different from the religious musings of other civilizations that are often called *philosophy* because its history was a progressive change in culture. Ontological scientists will not dismiss it as a historical accident because they will know from their explanation of evolution that culture evolves gradually in the direction of greater functional power, and their explanation of stages of evolution will enable them to explain Western civilization as a metaphysical stage in the evolution of life that follows the stage during which reflective subjects evolve. That stage is represented on Earth by non-Western societies.

It began in ancient Greece, and ontological scientists will notice that belief in the possibility of perfect knowledge preceded metaphysics. An inquiry by pre-Socratic philosophers was the context in which Plato introduced metaphysical arguments, and ontological scientists will trace their origin to them. The pre-Socratic philosophers were naturalists who set out to discover the *first cause* of the natural world. It would be perfect knowledge because the first cause would explain everything found in the natural world. In the end, the pre-Socratics agreed that substances are the first cause. Substances explain what is found in the world by constituting it, and since they explain what is found by its existence, they are *ontological causes* in our sense. Knowledge of the nature of the substances constituting everything in space and time would be a complete explanation for naturalists because they believe what exists is the natural world.

Though the pre-Socratics never agreed about the nature of those substances, their discovery about the nature of the first cause set a high standard for explanations that claim to be complete. It is what forced metaphysicians to recognize that their attempts to show that Reason knows Reality behind Appearance were failures, labeling philosophy the *dumping ground of unsolvable problems*. Descartes' problem of mind-body dualism is one such problem, and though its solution promises to cut the Gordian knot of philosophy and solve its problems, these days, no one expects that to happen.

That is, however, what *Naturalistic Reason* predicts ontological scientists will do when they trace their ancestry, by way of physical science, to Descartes' false belief about mathematics. The belief in intuitionistic Reason was a premise of his metaphysical argument, and assuming that all of consciousness is Appearance, Descartes argued that Reason knows Reality behind Appearance. But once natural philosophers had explained how animals perceive the natural world, intuitionists were forced to recognize that phenomenal appearances are part of the knower, and metaphysicians confronted another problematic ontological dualism. One was the Reality behind perceptual phenomenal appearances, and the other was the Reality behind phenomenal appearances of all kinds. Descartes used the clear and distinct ideas of mathematics to describe the essential nature the external world, so he believed that it was made up of material substances that exist outside one another in three dimensions leaving no voids (or *plenum materialism,* which he called *extension*). He used his famous argument, called the Cogito—*I think, therefore I am*—to know the Reality behind ideas of all kinds, and this was a description of the illusion of intuitionism. *I think* refers to ideas that are known by their immediate presence, and since there had to be a subject to whom they were present, *I am* refers to a substance distinct from extension. He called it *mind*, and it had a unity just opposite to the divisibility of extension because every quale in any configuration of qualia was immediately present to a single subject—including the clear and distinct idea called the *Cogito*. This discovery about the dualistic nature of the Reality behind Appearance posed an ontological problem that could not be solved in a way that passed the pre-Socratic test of ontological coherence. Descartes had to believe that mind and body interact with one another, and they were substances with opposite essential natures. Change occurs in a body because it is made of parts that can move relative to one another, while change occurs in mind because the immediate presence of ideas causes beliefs and choices about how to act, and thus, there was no way to explain how they jointly determine what happens in the natural world.

This is the origin of the *belief in mind*, and since ontological scientists will discover its cause by tracing the origin of ontological science, via physical science, to Descartes' belief that mathematics is known by rational intuition, they will have solved his problem of mind-body dualism. The solution implies that the belief in mind is false. The brain is what guides the body's behavior, and the mind is just a *phantom Reality* that metaphysicians conjure up when they are representative realists. Besides the external world behind objects presented by their faculties of perceptual intuition, Reason knows Reality behind phenomenal appearances of all kinds, so metaphysicians believe in a substance that cannot be part of the natural world. The apparent indubitability of the *Cogito* to language-using mammals explains why the belief in mind is so widely shared even though mind is not a substance.

This solution to the problem of mind-body dualism implies that the mind is identical to the brain because it shows how the illusion of intuitionism causes the false belief in mind. But since this solution is entailed by the explanation of Western civilization as a metaphysical stage of evolution that follows the stage at which reflective subjects evolve, it is a discovery of ontological science. Hence, the identity of mind and brain is a theoretical identification by ontological science in the same sense that the discovery of the identity of water and H_2O was a theoretical identification discovered by physical science. Science and philosophy are unified by the empirical discovery of mind-brain identity, and since that enables ontological science to explain everything found in the study of Appearance as well as everything found in the study of nature, it cuts the Gordian knot of philosophy. Thus, the first cause is spatio-materialism, and ontological science turns out to be the perfect kind of knowledge that the pre-Socratics hoped to discover.

This is the Second Enlightenment, but it is knowledge of the Good as well as the True. Historically, what enabled intuitionistic metaphysics to pass the Socratic test of practical wisdom was a marriage of Plato's dualism of Being and Becoming with the Christian belief in a God outside space and time who created the natural world. God's purpose in creating the world was an explanation of the nature of what is good that could cause Rational beings to choose to be moral even when it was contrary to self-interest or strong bodily desires. But it depended on another illusion, an illusion inherent in reflection rather than consciousness, called the *illusion of purposiveness*. As *Chapter 6* shows, the recognition of both illusions enables ontological scientists to explain Western civilization as the metaphysical stage of evolution in a way that implies not only that the mind is identical to the brain but also that God is identical to the natural world. This is how spatio-materialism is a first cause that explains what we ought to do as perfectly as it explains what we ought to believe. In both cases, it is what corresponds to Reality.

This sketch of how ontological scientists will explain everything found in the study of Appearance as well as everything found in the study of nature has taken it for granted that evolution can be explained as a series of inevitable stages that lead to the existence of beings like us. More details about that explanation are given in *Chapter 4*, but even at this high altitude, one aspect must be mentioned to show how ontological science becomes naturalistic Reason.

The discovery that there are geometrical efficient causes as well physical efficient causes is what makes it possible to reduce all the branches of science to spatio-materialism, and the ontological explanation of evolution as a series of inevitable stages depends on recognizing that they are caused by a series of levels of organization in evolving geometrical causes. If life evolves from nonlife at the first stage, reflective subjects evolve from non-reflective language-using mammals at the eighth stage (counting both major and minor stages), and metaphysical arguments are exchanged at the ninth stage. In the final three stages, spiritual organisms are the form of life that evolves, and the geometrical causes that evolve are the linguistic representations that members of those groups exchange. The grammatical structures of the sentences used identify the first two levels of linguistic organization, but the metaphysical level of linguistic organization is not so obvious. The first two linguistic levels depend on faculties of imagination that can be identified in the structure of the mammalian brain, but the third level depends only on exchanging arguments that describe parts of consciousness as objects of intuition. Reflective subjects use psychological sentences to represent psychological states as part of the processes by which they determine beliefs and behavior, and metaphysical arguments are a higher level of linguistic organization because they assume that the immediate presence of phenomenal properties are the cause of beliefs and behavior. That separates phenomenal intrinsic properties of a bit of matter that helps constitute psychological states from the psychological states that actually cause beliefs and behavior, and that is the level of linguistic organization that causes the metaphysical stage of evolution. But notice that although consciousness is part of the level of linguistic organization that causes the metaphysical stage of evolution, it is not the kind of efficient cause it seems to be. The immediate presence of phenomenal properties is not what causes knowledge of them. What causes the metaphysical stage is the illusion of intuitionism, that is, the assumption that their immediate presence is a cause that is presupposed by metaphysical arguments. For example, as we have seen, the *Cogito* describes the illusion, and since the illusion depends on the unity of consciousness, intuitionists believe that the immediate presence of phenomenal properties is a substance whose unity is incompatible with the divisibility of matter in the natural world. But mind is a phantom Reality, and since consciousness is just the object referred to by this misleading description of causes of behavior, it plays a strictly passive role in causing the belief in mind, and that is what I mean by calling consciousness a *pseudo-efficient cause*. The explanation of mind is fully compatible with epiphenomenal panpsychism. The only efficient causes in the spatio-material world are physical and geometrical causes.

When ontological scientists trace the origin of ontological science by way of physical science to the metaphysical stage of evolution, their explanation of the natural world will be complete. But since consciousness is just a pseudo-efficient cause, a final step is required for scientists to use ontological science to explain what they find in the natural world.

Since scientists are conscious, they will understand Descartes' metaphysical argument, and so they will refer to their own consciousness as something different from their brains. But ontological scientists use naturalistic imagination to show how substances entail everything that holds necessarily in the natural world, so they will use naturalistic imagination to explain their own understanding of Descartes. Since they know how consciousness is part of the natural world, they will know that there is an illusion inherent in consciousness, and so they will explain their own consciousness from outside consciousness. Thus, when they explain Western civilization as the metaphysical stage, they will know that consciousness plays the role of a pseudo-efficient cause in helping cause the metaphysical stage, and they will *discount the illusion of intuitionism* in themselves. Thus, they will *find themselves* knowing Reality behind Appearance. But they will know Reality in a different way from intuitionists because they will not use intuitionistic Reason. Instead of using Appearance, or the immediate presence of complex phenomenal properties, to infer the nature of Reality, they will discount the illusion of intuitionism. (See Figure 2 and compare it with Figure 1.) But they will have to refer to consciousness in order to understand the nature of Reality because discounting the illusion of intuitionism will be a necessary step in using ontological science to know Reality, and so they will recognize that ontological science knows Reality behind Appearance. That is the kind of cognitive power that metaphysicians called *Reason*. But since they are naturalists who use the empirical method to know Reality, ontological scientists will insist that their cognitive power be called *naturalistic* Reason. Thus, ontological science turns into naturalistic Reason.

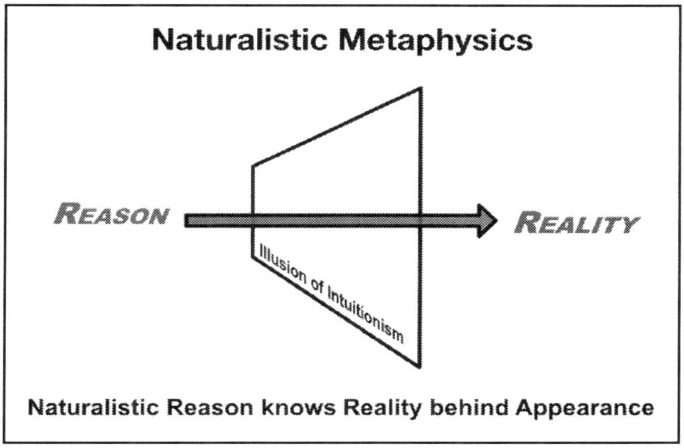

FIGURE 2 - NATURALISTIC METAPHYSICS

We may, therefore, be on the brink of a scientific revolution that gives us the perfect kind of knowledge to which philosophy has aspired ever since ancient Greece. *Chapter 1* has shown why we can expect to complete the Enlightenment that began in the 18th-Century, and *Chapter 2* has shown why it is not implausible to expect a Second Enlightenment to give us the perfect knowledge to which philosophy has long aspired. In the former, ontological causes explain the truth of *mathematics*, while in the latter, ontological causes explain the truth of *metaphysics*. Though both discoveries are plausible, the second is more surprising than the first. The first merely completes the study of nature begun during the Enlightenment. But since the second discovery comes from tracing the origin of ontological science to intuitionistic metaphysics, all the philosophical problems that have accumulated in the study of Appearance are solved by explaining Western civilization as a distinct stage of evolution that follows the stage at which reflective subjects evolve.

It is easier to dismiss the prediction of the Second Enlightenment than the prediction of the First Enlightenment because the belief that Western civilization is a stage of evolution that follows the stage represented by other civilizations on Earth contradicts multiculturalism and flies in the face of the rising tide of endarkenment. But the payoff of daring to know is also greater. It is what those who are nostalgic about the Enlightenment are looking for, whether they know it or not. It is a revolution in culture that will draw everyone into agreement on basic issues about what to believe and what to do because they will agree that what is true depends on correspondence to the Reality discovered empirically by science. And since this is a revolution that can spread by cultural diffusion, it can spread around the globe peacefully.

The prediction is that a scientific discovery that transforms culture in this way is not only inevitable but also imminent. *Part One* of *Sapere Aude* shows the unity and completeness of naturalistic Reason. But the trilogy lays the argument out in detail, and it may be complete enough that, if it is sound, it will even cause this revolution in culture. But it contains so many technical details that most readers will want to know more about crucial steps leading to naturalistic Reason before reading all three volumes. Thus, *Part Two* offers a simple and accurate executive summary of the argument, though at the end, it also goes beyond the trilogy by speculating about the very long-term future of life on Earth.

Part Two

Summary of the Argument

Sapere Aude is a summary of the argument for believing that we are on the brink of a revolution in which science becomes Reason knowing Reality behind Appearance. The three volumes of *Naturalistic Reason* are represented in Figure 3.

FIGURE 3 - NATURALISTIC REASON

The triangle at the bottom of Figure 3 stands for *Volume I*, which predicts the *Unification of Physics.* It shows how a discovery about the nature of space will solve the puzzles of physics by reducing it to ontology. The arrow pointing to the circle labeled *Science* stands for *Volume II*, which predicts the *Unification of Science.* It shows how the reduction of physics to ontology reveals an efficient cause not recognized by physics whose way of working together with physical causes enables scientists to reduce all the regularities studied by the specialized sciences to interactions of space and matter, completing the First Enlightenment. Finally, the two arrows pointing toward the circle labelled *Philosophy* stand for *Volume III.* It predicts the

Unification of Science and Philosophy by showing how a combination of contributions from *Science* and *Ontology* provide the foundation for explaining Western philosophy as another inevitable stage in the evolution of life on Earth, which culminates in naturalistic Reason. The following chapters summarize all three volumes, and *natReason*, at the center of the diagram, stands for the entire path from the discovery in physics about the nature of space to the transformation of ontological science into naturalistic Reason. But since the Good is part of the True, this argument has implications for practical Reason as well as theoretical Reason. The final chapter of Part Two of *Sapere Aude* defends a practical implication that is only hinted at in the trilogy.

Chapter 1

Unification of Physics (Volume I)

Modern physics is in a bind. On the one hand, it has had unparalleled success in discovering mathematically formulated laws of nature. On the other hand, it came up with laws of physics describing two basically different kinds of regularities, and not only is each puzzling in its own right, but their combination is also puzzling because they are mathematically incompatible. No mathematically formulated law of nature can be found that describes the relation between the two puzzling regularities. Physicists take pride in their achievement, but the lack of progress in solving these problems during the past few decades has them worried. A consensus is growing that physics is missing something basic. But what it is missing is so basic that physicists are blind to it. The basic branch of science must explain *what exists*, not just *what happens.*

The laws of quantum physics describe regularities on the smallest scale, while Einstein's general theory of relativity describes a gravitational regularity that holds on the largest scale. Quantum physics is puzzling because it had to give up the classical assumption that laws of physics are deterministic and accept probabilistic laws as basic. Gravitational physics is puzzling because it had to give up Newtonian absolute space and time in favor of spacetime, a 4-D entity in which time is just a time-like spatial dimension. These implications are puzzling because in neither case can what corresponds to the laws be understood in *naturalistic imagination*, the cognitive faculty built into our brains that enables all mammals to think about the relations of objects in space and predict how motion changes them. What is worse, the two basic theories of modern physics are inconsistent. Probabilistic laws of quantum physics describe a non-local regularity, called *entanglement,* in which the measurement of one particle seems to affect what is measured in another particle located far away faster than the speed of light. According to Einstein's special theory of relativity, that is not possible. But that problem seems minor by comparison with the most pressing theoretical problem facing modern physics. The equations used to formulate the laws of quantum and gravitational physics are so radically different from one another that no one can find equations that describe both regularities at once. Physicists must look to the Big Bang to find a situation in which the energy levels described by the two kinds of forces are even comparable.

But the mathematical nature of the disparity between the two branches suggests what physics is missing. Judging by its practice, the method that has made physics so successful is empirical lawism. It assumes that the most basic possible empirical knowledge about the natural world is mathematically formulated laws of nature. Since they can be confirmed by precise measurements of quantitative prediction, there is no doubt that physics is onto something basic about the natural world. Hence, the attempt to formulate a single law of nature mathematically that describes all the basic regularities at once is called the search for the *holy grail.* It is dubbed the *TOE,* or *Theory of Everything.* But even if it were possible, it would be a pyrrhic victory if it described regularities, like those described by the two basic physical theories, that could not be understood in naturalistic imagination.

The method of physics is what hides the discovery that will solve all the puzzles of modern physics. Physicists do not call their method *empirical lawism.* Theoretical physicists just take it for granted that laws of physics must be formulated mathematically because they know that what keeps them honest are experiments in which measurements can either confirm or falsify them. Precise measurements are typically so difficult that they are carried out by other physicists, called *experimental physicists.* What physicists have in common is the use of mathematics as a language for describing regularities, so both kinds of physicists take it for granted that mathematics is known independently of perception. As we have seen, there are physicists who wonder at the "unreasonable effectiveness" of mathematics in discovering laws of physics. But they do not consider the possibility of solving this mystery by discovering a theory that explains the truth of mathematics by its correspondence to the natural world.[6] That *is* what solves this mystery—and all the other puzzles of modern physics. But their method keeps them from being able to imagine the discovery that would show that mathematics is true by correspondence.

---

[6] Except for Max Tegmark (2014) *Our Mathematical Universe: My Quest for the Ultimate Nature of Reality,* who argues that the natural world is nothing but mathematics, a theory like the one defended by Pythagoreans in ancient Greece.

Empirical lawism assumes that the basic branch of science discovers laws of nature, and that is not necessarily true. As we have noticed, the pre-Socratic philosophers in ancient Greece had another way of explaining what is found in the world. The world includes everything that exists, and as naturalists, the pre-Socratics assumed that the natural world is what exists. They were looking for its *first cause*, that is, a cause that would explain everything found in the natural world. They concluded that it must be *substances* that constitute it. Substances were assumed to be self-subsistent entities that endure through time, and with powers expressed by interactions with one another, they could be used to explain the kinds of change found in the natural world. *Ontology* means the study of what exists most basically, and substances are fittingly called *ontological causes* because they explain what is found in the natural world by causing its existence, that is, by constituting it. As the first cause, substances constitute everything found in the natural world. Though there were several generations of pre-Socratic philosophers, they never agreed about the nature of those substances. Since they could not use mathematics to describe quantitative regularities precisely, their way of choosing between alternative theories was not exacting enough to learn much more about the substances constituting the natural world.

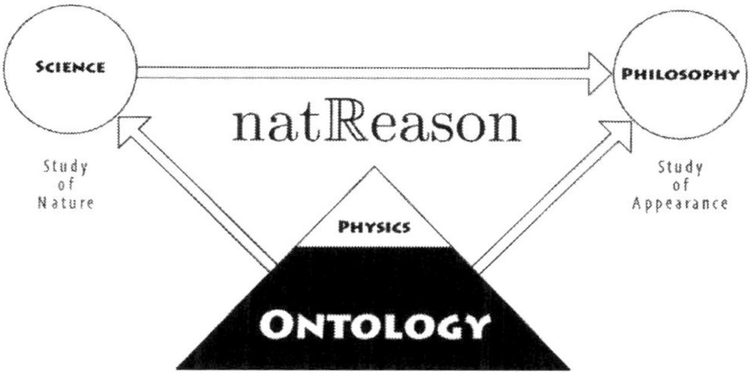

FIGURE 4 - UNIFICATION OF PHYSICS

However, we are in a different position because modern physics has used mathematics to discover basic quantitative regularities about change, and we can use the empirical method to discover the kinds of substances that explain the laws of physics. See Figure 4. This is the method I call *empirical ontology*. It agrees with the pre-Socratics about substances being the first cause, so it assumes that the deepest possible empirical knowledge about the natural world is about the substances constituting the world. The discovery of their natures is what will solve the puzzles of modern physics and turn science into naturalistic Reason. But it depends on explaining not only the truth of the laws of physics, but also the truth of the mathematics used to formulate them.

(1) Nature of Mathematical Truth

Ontology will be able to solve the puzzles of physics because there are substances whose kinds solve the mystery about the unreasonable effectiveness of mathematics in discovering laws of physics. If both space and matter are substances, and if they constitute the natural world, then ontological causes explain why the confirmation of laws of physics is so compelling. All the regularities generated by interactions of space and matter are quantitative.

Space is a one-of-a-kind infinite substance with a Euclidean geometrical structure. Since it is self-subsistent, its existence does not depend on matter, and likewise, matter has an intrinsic quantity that does not depend on space. Bits of matter coincide with parts of space, and they are distinct bits when they coincide with different parts of space or when they coincide with the same parts of space in different ways. Each bit of matter has a definite quantity, and differences in species of bits of matter are also quantitative because species are defined by the geometrical structure of their coincidence with space. This is the general kind of ontological theory denoted by *spatio-materialism*, and it implies that change of all kinds is change in the spatiotemporal (geometrical) structures of the coincidence of space and matter. Hence, all the regularities that interactions of space and matter can generate are, at bottom, about changes in quantitative properties.

Spatio-materialism was not among the ontological theories considered by pre-Socratics. They did not even have the concept of space as something different from matter. That came from Newton's need for something to correspond to the Cartesian coordinate systems that he used to formulate his laws.[7] Nor was any reason to believe that regularities are quantitatively precise prior to the discovery of mathematically formulated laws of nature. But in the spatio-material world, all regularities are quantitative, and that is why physics was so successful. History shows that physics began with the use of equations to describe quantitative regularities, and as we have seen,

---

[7] The ancient atomists believed in atoms and the void, not atoms and space. The void was explained as what existed between atoms, presumably moving out of their way like a fluid as they moved.

Kant's constructivist explanation of the nature of Reason allows physicists to believe that mathematics is *a priori*. Thus, mathematics seems to be just a language in which laws of nature could describes the most basic regularities. But without a mathematically minded creator of the natural world, naturalists cannot explain why mathematics is so unreasonably effective in discovering laws of physics, and its explanation is what is missing in physics. When empirical ontologists infer spatio-materialism as the best explanation of the unreasonable effectiveness of mathematics in discovering laws of physics, they will explain not only the unprecedented success of physics, but also what causes its intractable puzzles. To see how, consider the analogy between empirical lawism and a *mathoscope*.

To use mathematics to formulate laws of nature is like using a microscope or telescope, except that its lenses are coordinate systems and its images are equations that use those lenses to refer to the natural world and describe regularities about change. The regularities are generated by interactions of space and matter, and since images in mathoscopes are quantitatively precise, they entail quantitative predictions. Measurements of them are a compelling way to confirm laws of nature, and since their confirmation is convincing evidence of the existence of laws of physics, physicists are naturally inclined to believe that the images in their mathoscopes describe what exists most basically in the natural world.

The discovery about space on which this prediction of an imminent cultural revolution is based is not merely that space is a substance. It is the discovery that space is a substance that *inter-acts* with matter, the other kind of substance helping it constitute the natural world. If space were just a container in which particles moved and interacted, as Newton assumed, the use of coordinate systems to refer to the natural world would enable physics to use equations to describe *completely* space's role in helping generate quantitative regularities. All laws of physics would describe regularities that could be explained ontologically as expressing powers of the material substances contained by space. But if space is a substance that inter-acts with the bits of matter coinciding with it, each substance acts on the other, affecting its powers, and both substances are part of each particle. When regularities about change express powers of both space and matter, it is possible for their interactions to generate quantitative regularities that cannot be described completely by equations using coordinate systems to refer to particles.

This is what causes intractable puzzles in modern physics, albeit in different ways in quantum and gravitational physics. In both cases, it depends on the belief that mathematics is known *a priori*. Since that is an infection of physical science by intuitionism, spacetime is a *phantom Reality* in the same sense as mind. The interpretation of the probabilistic character of quantum laws as indeterminism is another phantom Reality. In other words, ontological science solves the puzzles of modern physics in much the same way it solves the problem of mind-body dualism: by reducing phantom Realities to interactions of space and matter.

Volume I describes powers of space and matter that will enable ontologists to reduce the laws of classical and modern physics to interactions of space and matter. They are complex, as suggested later in this chapter, and *Volume I* describes them in quantitative detail. But spatio-material interactions are ontological mechanisms, and since naturalistic imagination enables us to understand the geometrical structure of space, it also enables us to understand interactions of space and matter. Thus, ontological mechanisms are what solve the puzzles of physics.

To avoid misunderstanding, let me emphasize that the discovery of spatio-materialism is not just an interpretation of laws of physics. *Interpretations* accept the truth of physical laws and, then, describe what *must* exist in the natural world for them to be true by correspondence. The laws are puzzling because what must exist cannot be understood in naturalistic imagination. But spatio-materialism is not an interpretation of laws of physics. Ontology is a more basic science than physics. *What exists* is prior to *what happens* because *what happens happens* to *what exists*. Ontology *assumes* that substances are the first cause, and since that is to assume that laws of nature can be explained by substances and their powers, it rejects *empirical lawism* in favor of *empirical ontology*. And when ontologists infer that space and matter constitute the natural world as the best explanation of the unreasonable effectiveness of mathematics in discovering laws of physics, they will be able to infer more specific powers of space and matter as the best ontological explanation of the regularities described by laws of physics.

If spatio-materialism explains the appropriateness of mathematics in formulating laws of nature, the notion of a mathoscope explains not only the success of physics but also the cause of its puzzles. Since its lenses are coordinate systems that correspond to space, they filter out some roles that space plays in helping matter generate regularities about change. *Volume I* shows how the omission of space's role makes *interpretations* of physics puzzling. The probabilistic laws of quantum physics are images of regularities generated by deterministic ontological mechanisms on the smallest scale, and the images are *blurry* because the equations used cannot describe those regularities completely. A different kind of mathoscope is used in modern gravitational physics, and since the equations used to formulate Einstein's general theory of relativity are *aberrant* images, it seems that Newtonian absolute space and time must be replaced by spacetime.

The mathematical disparity of quantum and gravitational physics is represented by the inability to combine the images presented in these two kinds of mathoscopes as an image presented in a single mathoscope. These two phantom Realities cannot be merged into a single phantom Reality. But the reduction of physics to ontology will show how mathematics is such a tool, and since it will explain the relation between the laws of quantum and gravitational physics, it will unify physics ontologically. But as we shall see, that is not the most far-reaching implication of the discovery about space predicted in *Volume I*.

(2) Reduction of Quantum Physics to the Inertial system

When ontologists show how the inertial system generates the regularities described by laws of quantum physics, they will discover that it generates other regularities as well, and they are the foundation for the predictions defended in *Volume II*. Though this brief sketch of its basic structure cannot explain all the quantitative aspects of quantum laws, it will show how to picture the inertial system as an ontological mechanism that explains the regularities they describe.

Except for space, nothing exists in the spatio-material world but matter, so matter helps constitute all the energy and mass covered by Einstein's famous equation: $E = mc^2$. Energy can be explained ontologically as the quantity of matter, and mass can be explained as a measure of the quantity of matter in terms of inertia (and gravitational attraction), since energy and mass are still proportional (though the relation is more complex). All the particles, particle motion, and particle interactions described by the laws of quantum physics (including all the particles in the *Standard Model*) are constituted by the inertial system. (See Figure 5.)

Ontologists will discover that the inertial system is made up of two parallel parts: the electromagnetic system and the mechanical system. Charged particles are constituted by both systems, and working together, they explain interactions of electrons, protons, and the nuclei of other atoms as a flow of matter in the electromagnetic system, called *acceleration fuel*, which is measured as *action* in physics. The particles and their motion are both constituted by bound matter *and* the parts of space with which they coincide in both systems. (See Figure 5.) The mechanical system gives particles precise locations in space, and the sizes of all its elementary parts are measured by a unit of action, called the *quantum*, which is equal to Planck's constant. The rest energy of a particle is determined by the number of *temporal* quantum actions of its movable matter per unit of time in the mechanical system; its velocity is a number of *spatial* quantum actions per unit of distance that a particle with its rest matter has in motion; and its kinetic energy is determined by the number of *temporal* quantum actions of motor matter per unit time in the mechanical system. The electromagnetic system gives particles an electric charge that is surrounded by electric field matter when the charge is at rest, and when the charge is in motion, it is also surrounded by magnetic field matter. The quantity of field matter depends on the densities of its coincidence with space, and its quantitative properties are also determined in the inertial system by the quantum structure of the mechanical system, since bits of bound matter always have three times as much electromagnetic matter as mechanical matter.

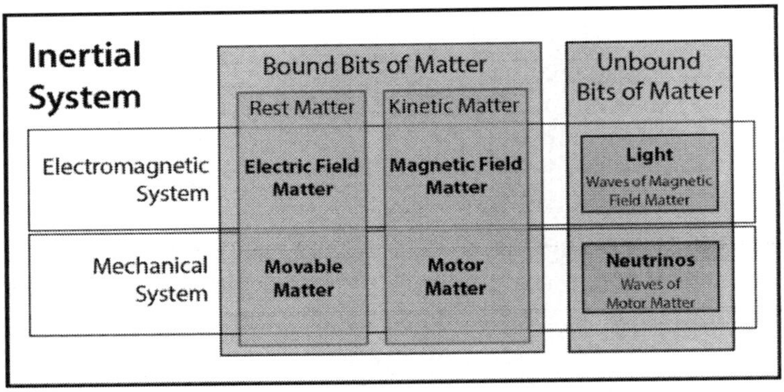

FIGURE 5 - SPECIES OF MATTER

The inertial system is the ontological mechanism that generates all the regularities described by laws of quantum physics, and this brief sketch suggests the degree of its complexity. According to the *fueling model of forces*, flows of field matter between charged particles in the electromagnetic system enable them to change one another's motion by changing their kinetic energy (or the number of temporal quantum actions per unit time helping constitute their motion in the mechanical system), and so field matter is the acceleration fuel that mediates interactions of particles. In physics, its flow is measured as *action*.

The laws of classical physics are implicit in the laws of quantum physics, and while the mechanical system explains the regularities described by Newton's laws of motion (with quantitative aspects explained by numbers of quantum units of action), the electromagnetic system explains the regularities described by the laws of electromagnetism (both Maxwell's equations and the Lorentz force law). This explains the regularities about the motion and interaction of particles moving independently of one another, called *free particles*. Since they are described by basic laws of physics, they determine what happens to free particles, and so this is how the inertial system explains what are called *physical (efficient) causes*.

Indeed, since flows of acceleration fuel between charged particles are measured as action, ontologists will discover that the inertial system explains the puzzling *principle of least action*. According to that principle of classical physics, when the change in a closed system between its beginning and ending states has the least action, the particles in the system move and interact in conformity with the laws of physics. Action is a measure of flows of acceleration fuel between field matter and kinetic matter, and since the fueling model of forces in the inertial system implies that the flow from field matter to kinetic matter required to accelerate a particle varies directly with its velocity, the inertial system explains the principle of least action ontologically. (In a nutshell, a charged particle moving through field matter with a constant density at a higher velocity will accelerate at a constant rate because at higher velocities it acquires more kinetic matter from field matter per unit time along its path.)

The inertial system explains electromagnetic waves, such as light, as magnetic field matter that has been freed from helping constitute charged particles in the electromagnetic system and is carried across the electromagnetic system at the speed of light. In quantum physics, waves of magnetic field matter carried across the electromagnetic system at the speed of light are called *photons* because they seem to have a particle-like nature. But that will be explained ontologically as just an appearance caused by how magnetic field matter waves interact with charged particles because of the mechanical system. (Similarly, waves of motor matter that have been freed from helping constitute charged particles in the mechanical system and are carried across the mechanical system at the speed of light explain the puzzling quantum particles called *neutrinos*.)

Even the constant motion of a particle is a step-like change. Spatial quantum units of action have endpoints with precise locations in space and time, which are called *atomic events*. Since the *linear unit* occupied by a spatial quantum action cannot change, interactions can change the velocity of a charged particle only at endpoints when a flow of acceleration fuel changes its quantity of kinetic matter. This suggests how to picture what I call the *quantum structure of the mechanical system.* If we picture flows of field matter as conserving energy, the electromagnetic system determines quantities of kinetic matter of particles in the mechanical system, and its quantum structure conserves momentum at each and every atomic event. (What I call the *endpoint-orientation mechanism* is an ontological mechanism in which adjustments of endpoints of spatial quantum actions are balanced by changes in precise orientations of particle spin.)

This picture of the inertial system suggests how it generates regularities that solve other puzzles about quantum regularities. For example, the quantum structure of the mechanical system is responsible for the puzzling quantum property called *entanglement.* It is demonstrated dramatically by Bell's inequality, the famous probabilistic prediction about the measurement of one particle affecting the measurement of another particle far away faster than the speed of light. The laws of quantum physics entail this *non-local property,* and it is puzzling because particles are represented in the quantum mathoscope as having properties that they carry with them as they move across space, called *local properties.* But as ontologists will explain it, this puzzle is caused by the quantum mathoscope filtering out the role of space in the inertial system helping matter generate quantum regularities. Entanglement is caused by an interaction of space and matter in the quantum structure of the mechanical system, and since space is a single substance, it interacts with all the bits of matter coinciding with it everywhere at once. This does not violate the speed limit that Einstein's special theory of relativity imposes on one particle affecting another at a distance. Indeed, the inertial system also explains that regularity.

(3) Reduction of Gravitational Physics to the Gravitational System

While the blurriness of the image in the quantum mathoscope hides the quantum structure of the inertial system, ontological scientists will find that the image in the gravitational mathoscope is *aberrant* because its lenses make space seem to be spacetime. In the spatio-material world, the gravitational regularity is generated by the gravitational system, and to explain spacetime as an aberrant image, ontologists will, first, have to discover the ontological mechanism by which the inertial system generates the regularity described by Einstein's special theory of relativity.

That regularity is described by the *Lorentz transformation equations,* and the main premise that Einstein used to derive them was the *principle of relativity*, the assumption that the laws of (quantum) physics hold the same way in all reference frames moving at a constant velocity, regardless of their relative velocity. Assuming that light has the same velocity in every such reference frame, he derived equations that describe a regularity about the relation between reference frames moving relative to one another at speeds approaching that of light. Since they were equations that transformed the coordinates used in one reference frame to obtain those used in the other, they were an *indirect description* of the regularity in the spatio-material world, and the regularity was puzzling because it implied that what observers in one reference frame would find in the other reference frame are measuring rods that are contracted in the direction of relative motion and clocks that are slowed down, both at the same (*Lorentz*) rate as a function of relative velocity. The Lorentz transformation equations imply that reference frames cannot move faster than light because, at that relative velocity, the others' measuring rods would have no length at all and their clocks would have stopped entirely.

The regularity about the relation between so-called *inertial reference frames* that Einstein described is puzzling because naturalistic imagination does not enable us to understand how two measuring rods passing one another could both be shorter than the other. Nor how two clocks could both tick off seconds more slowly than the other. To be sure, it is not physically impossible. Observers on each reference frame can explain the puzzling measurements made by observers on the other reference frame as *mis*-measurements that are caused by using contracted rods and slowed-down clocks.[8] But this way of explaining away what the other observer measures is also symmetrical, and since empirical lawism leads physicists to accept mathematically formulated theories that are well confirmed, they believe that the symmetry is what really exists. That is why Einstein's discovery is called a theory of *relativity*.

Einstein described this symmetry in 1905, and it is not necessarily basic. The equations used to describe it are called the *Lorentz* transformation equations because H. A. Lorentz had already used them in 1905 for an entirely different purpose. He was trying to explain why the Michelson-

[8] This is possible because the other two premises of Einstein's special theory of relativity are the assumptions (1) that the velocity of light, *c*, is the same in all inertial reference frames and (2) that simultaneity is measured on each in the way that it would be measured at rest in absolute space.

Morley experiment in 1887 had failed to measure the velocity of Earth relative to space. They had assumed that light has a constant velocity in reference frames at absolute rest, and Lorentz found that the null outcome of their experiment could be explained if the absolute velocity of material bodies caused deformations in them, including length contraction and time dilation, both at the Lorentz rate, the same rate that Einstein had derived in his special theory. Lorentz had started with a reference frame at absolute rest, and he used transformation equations to describe its relation to a moving reference frame because he wanted to describe *all* the deformations that bodies suffer in terms of each other, that is, all at once. I call them *Lorentz deformations* because *Volume I* predicts that ontological scientists will follow Lorentz and explain them as regularities generated by the inertial system. Since Lorentz's premises were the laws of electromagnetism, his argument was basically mathematical, and it was a convoluted argument with problems of its own. But minor refinements of the quantum structure of the mechanical system will enable ontological scientists to use the inertial system to explain the Lorentz deformations ontologically, as described in *Volume I*. This is the reduction of Einstein's special theory of relativity to ontology.

As a classical physicist, Lorentz used equations that depend on coordinate systems for their references to the natural world, and that is the kind of mathoscope that quantum physicists use to describe regularities about change. The difference is that Lorentz used transformation equations to describe the *Lorentz deformations*, and that was to describe them indirectly. By using the inertial system to explain them, however, ontological scientists will discover the role that space plays in generating them. By discovering what was omitted by Lorentz' way of using the quantum mathoscope, ontological science will bridge the gap between quantum and gravitational physics.

Einstein's general theory of relativity is the basic law of gravitational physics, and as I have suggested, his gravitational mathoscope is different from the quantum mathoscope. The Image in his gravitational mathoscope is the Einstein field equations, and though they are transformation equations, they are fundamentally different from those that Lorentz used because they depend on another way of representing the Lorentz transformation mathematically. Minkowski found that it could be represented by a *spacetime* diagram, and the equation that Minkowski used to construct his diagram is what Einstein used as a coordinate system in general relativity. The relation between space and time are represented geometrically in two dimensions by using one spatial dimension to represent the direction in which one reference frame is moving relative to the other and using the other dimension to represent time. Units are described in such a way that the diagonal at 45^0 between the two axes represents the pathway determined by the speed of light, and Minkowski showed how his equation for a hyperbola in this diagram could be used to represent the relations between inertial reference frames moving at all possible velocities relative to one at the origin. This turned the Einsteinian relativity of spatial and temporal distances in each pair of inertial reference frames into an absolute truth about them that observers on all possible inertial reference frames could accept. But this mathematical absoluteness came at a cost. Different inertial reference frames at any point could be represented as different degrees of rotation of their space-like and time-like axes relative to the axes of Minkowski's spacetime diagram. If that describes what really exists in the natural world, then all moments in time are related to one another like points on a line, and every moment in the history of the universe exists in exactly the same sense as every other moment. This is called *eternalism* about time (and existence), and it contradicts our understanding of time in naturalistic imagination. Naturalistic imagination uses the passage of time to represent changes in relations between different moments over time. That is, when we use naturalistic imagination, we take it for granted that the existence of the present is what makes the present moment different from moments in the past, which no longer exist, and from moments in the future, which have yet to come into existence. We use changes in pictures as time passes to represent the passage of time.

Though Einstein based his general theory of relativity on his special theory, he introduced a new kind of mathoscope because he used the equation that Minkowski had introduced to describe his spacetime diagram as a 4-D coordinate system in which to formulate transformation equations that describe the acceleration of a reference frame in a gravitational field. Using complex tensor equations that had been developed to describe non-Euclidean geometries, general relativity attributed gravitation to a curvature of spacetime. The Einstein field equations entailed non-Newtonian predictions about gravitation, famously including the anomalous precession of the perihelion of Mercury, the bending of light rays passing through a gravitational field, and the slowing down of clocks at rest in a gravitational field. With the confirmation of these and other surprising implications, such as the expansion of the universe after the Big Bang, general relativity was accepted as the standard explanation of gravitation in modern physics.

Since Minkowski spacetime cannot be understood in naturalistic imagination, neither can general relativity. Ontological scientists will know that it is an *aberrant image* caused by the lenses in the gravitational mathoscope when they discover an ontological mechanism that generates the regularities on which the gravitational predictions of general relativity depend. The *gravitational system*, as I call it, is the all-inclusive ontological mechanism in the spatio-material world discovered by the reduction of physics to ontology. It is basically just a Newtonian gravitational field in which what is accelerated is the inertial system (which generates the Lorentz deformations as well as all the quantum regularities). Since the acceleration of the inertial system accelerates the matter that is interacting with space in it, its acceleration explains the acceleration of bound matter predicted by Newton's law. But it also gives the inertial system an inward velocity everywhere in the field that is equal to the classical escape velocity for a Newtonian field. Thus, light rays that are passing by a gravitating body are bent in its direction, and clocks at rest in a gravitational field suffer a Lorentz time dilation as a function of its inward velocity where the clock is located. Since bodies in orbit around a gravitating body have an outward velocity relative to the inertial system where they are located, they suffer a length contraction in the radial direction that is greater than their length contraction in the orbital direction. The net effect of the difference in these rates of length contraction is that the orbiting body does not travel as far outward in the gravitational field per unit time as Newton would predict, and since it falls faster than Newton would predict, its orbit precesses slightly in the direction of its orbit.[9] In sum, ontological scientists will discover that what corresponds to the curvature of spacetime is the acceleration of the inertial system at each point in a Newtonian gravitational field that causes Lorentz deformations in bound matter located there.

What enables the inertial system to generate Lorentz deformations is a minor refinement of the quantum structure by which the inertial system generates all the regularities described by laws of quantum physics, and as *Volume I* predicts, it will enable ontological scientists to explain the apparent expansion of the universe after the so-called Big Bang. What will replace the expansion of spacetime is a shrinking of bound matter (both the quantum

---

[9] But resistance to acceleration in the longitudinal direction increases at the cube of the Lorentz rate, and as *Volume I* explains the inertial system, the net precession of an orbit also depends on a regularity about how time dilation depends on the cubed Lorentz rate of longitudinal mass increase that even Lorentz did not recognize.

structure in the mechanical system and the charge sphere radius in the electromagnetic system). That will offer a new way of explaining puzzling astronomical phenomena now explained on the assumption that the universe began with the Big Bang with the assumption that little big shrinks are continually occurring in empty regions of infinite space. This shrinking of bound matter in the gravitational system will replace *dark energy*. The multiplicity of little big shrinks will explain the observations that have led cosmologists to believe that the expansion of the universe started to increase five or six billion years ago. And *dark matter* will probably be replaced by surprising implications of the inertial system that are required to explain quantitative aspects of laws of classical physics.[10]

(4) Mathematical Disparity of Quantum and Gravitational Physics

The most pressing problem of theoretical physics is the mathematical disparity between quantum and gravitational physics, and *Volume I* predicts that ontologists will discover that it is caused by failing to recognize that the mathoscope used to describe gravitational regularities on the largest scale is basically different from the mathoscope used to describe quantum regularities on the smallest scale. In both branches of physics, what a mathoscope filters out is a role that space plays in helping matter generate regularities about change. What the mathoscope in quantum physics filters out is the *inertial system*, including the ontological mechanism responsible for the Lorentz deformations, and what the mathoscope used in gravitational mathoscope filters out is the *gravitational system.* Since the inertial system is part of the gravitational system, ontologists will solve the mathematical disparity between quantum and gravitational physics by discovering how both kinds of equations describe regularities generated by the same all-inclusive interaction of space and matter.

Ontology is a way of thinking outside the mathematical box of physics, and solving the puzzles of physics will convince most physicists that ontology is more basic than physics. For example, it explains the *fine structure constant*, which is so puzzling that the physicist, Richard Feynman, said that all physicists should post it on the walls of their offices to remind them of what they do not know. Some may resist accepting empirical

---

[10] In brief, to explain the equations used to formulate Newton's laws of motion, ontologists will have to hold that bound matter in the electromagnetic and mechanical system systems both have an inertia equal to the inertia attributed to particles, and since that doubles the mass in the universe, gravitational attraction is quadrupled.

ontology on the grounds that there is no value in being able to understand in naturalistic imagination what corresponds to its mathematically formulated laws of nature. Empirical lawists can insist that everything depends on measurements that confirm quantitative predictions. But empirical lawism will compel them to accept ontology because its explanation of physics also entails measurements not predicted by laws of physics. For example, it offers a solution to the recently recognized G-2 (*G minus two*) puzzle. Though the Standard Model predicts the anomalous magnetic moment of the electron, it does not predict that property of the muon accurately, and the reduction of quantum physics to ontology shows how the difference could be caused. Similarly, the reduction of gravitational physics to ontology will show how Earth's velocity relative to space can be measured, and since that is inconsistent with Einstein's (special) principle of relativity, a measurement of absolute velocity will falsify the belief in spacetime. The cost of refusing to accept a unification of quantum and gravitational physics that solves the puzzles confronting both will be having to explain away ontological predictions of measurements that contradict laws of physics.

Chapter 2

Unification of Science (Volume II)

On the path to naturalistic Reason, however, the most far-reaching consequence of the reduction of physics to ontology is the discovery that matter has a *holistic power*. The holistic power of matter is entailed by the inertial system to which the laws of quantum physics are reduced, and it is the foundation on which the other two volumes of *Naturalistic Reason* predict further discoveries in science. The quantum structure of the mechanical system expresses a non-local power of space to interact with matter, and based on its intrinsic geometrical structure, it can give matter a holistic power that is expressed geometrically—in two ways. One is the foundation for predicting the unification of science in *Volume II*, and a further expression is the foundation for predicting the unification of science and philosophy in *Volume III*.

The first expression enables groups of particles bound together as parts of atoms (and material bodies composed of them) to move around independently and act on the world as a whole. Such *composite bodies* have unchanging geometrical structures, and since they act on the world as a whole, their geometrical structures impose constraints on what happens by physical causes. It is an efficient cause. But it is a kind of efficient cause that is not recognized in physics. *Physical causes* are entailed by laws of quantum physics describing regularities about the motion and interaction of particles, and physicists assume that they are responsible for all the regularities found in the natural world. But I call composite bodies with unchanging structures *geometrical causes* because they are just as basic as physical causes. They can jointly determine what happens, and discovering how they work together is what will solve the puzzles of science and complete the First Enlightenment.

Moreover, together with a further expression of the holistic power of matter, geometrical causes will enable ontological science to explain its own existence, and that will turn science into naturalistic Reason, completing the Second Enlightenment. This second expression of the holistic power of matter depends on matter having a phenomenal (or quality-like) way of existing in itself, so that when its coincidence with space gives it the right kind of spatiotemporal structure, its intrinsic property constitutes consciousness. As we shall see, since the basic structure of the mammalian brain is a geometrical cause that serves as a faculty of imagination, the constraint it imposes on the motion and interaction of particles throughout the brain explains consciousness as what it is like to be the acceleration fuel mediating their interaction. Thus, once geometrical efficient causes enable science to unify all the specialized branches, a further effect due to the phenomenal intrinsic nature of matter will enable science to explain how consciousness, as part of the natural world, helps cause the metaphysical stage of evolution. That is what solves the puzzles of philosophy.

(1) Ontological Reduction of Quantum Physics

Geometrical causes are composite bodies, and since they and their effects play such a big role in the natural world, it hardly seems possible that they are not already recognized in physics. Since composite bodies are the tools used by tool-making animals, it is obvious that their geometrical structures can constrain what happens: a key can open a lock, and a box can contain a gas. Something similar happens on the micro-level in living organisms when complex molecular machines, called *ribosomes*, transcribe (messenger) RNA into protein molecules. Composite bodies can have unchanging geometrical structures that constrain *what happens*, and it is proper to call them *geometrical* efficient causes because they impose geometrical constraints on *what happens by physical causes*. Physicists do not recognize geometrical causes because they assume that composite bodies are themselves consequences of physical causes. Physics is the basic branch of science, and since empirical lawism implies that physical causes are responsible for everything that happens, physicalists assume that composite bodies are reducible to physical causes.

To be precise, physicalism implies that physical causes are complete in the sense that they explain every event that can be explained. But not every event can be explained completely. For example, when a particle moves *freely*, that is, independently of other particles, the *Heisenberg uncertainty principle* implies that it is not possible to measure both its position and momentum with less than a small amount of uncertainty about the size of the quantum unit of action. Indeed, since the probabilistic character of regularities is basic to laws of quantum physics, it holds even when they are used to explain composite bodies. Thus, from nuclei and atoms up to all the bodies composed of them, the physical explanation of their geometrical structures depends on another puzzling quantum principle, the *Pauli exclusion principle.* It holds that it is not possible for more than one particle of the same kind to occupy the same *quantum state*, and by keeping track of all possible quantum states, physicists can explain the robustness of the geometrical structures of composite bodies by the strengths of the probabilistic bonds between particles making them up. Though physicists believed at the beginning of the quantum revolution that electrons in atoms orbited their nucleus, they now describe them as occupying *electron orbitals.* They are just *probability densities*, or the probability of finding an electron at each point in a region, and in the most complete physical explanation of atoms, probability densities are just waves described by the *Schrödinger wavefunction*.

This will turn out to be a blurry image of atoms in the mathoscope of quantum physics. The reduction of physics to ontology will reveal that interactions of space and matter in the inertial system generate completely deterministic quantitative regularities about how particles move and interact. As mentioned previously, the constant motion of a particle is a change of location punctuated by atomic events. Each step is a spatial quantum of action whose endpoints have precise locations in space and time, and since any change in a particle's motion depends on a flow of acceleration fuel, it can interact with the rest of the world only at endpoints. Momentum as well as energy are conserved at each atomic event by the endpoint-orientation mechanism. The quantum structure of the inertial system is what corresponds to the quantum property, *entanglement*, and since it expresses a power of space, all particles in the universe are entangled.

The image of the quantum structure of the inertial system in the quantum mathoscope is blurry because its lens filters out the details of this deterministic ontological mechanism. But as this regularity is described by the *Heisenberg uncertainty principle,* it is not possible to measure both the position and momentum of a free particle with an uncertainty less than a fraction of Planck's constant, and a similar blurriness holds for particles in atoms. However, as ontologists will discover, atoms are made up of particles bound together by cycles of interactions that can be repeated endlessly. Particles can accelerate only at endpoints of spatial quantum actions, and in atoms, atomic events conserve both energy and momentum because, as parts of cycles that can be repeated endlessly, they are related in space and time in a way that does not require any adjustments in endpoints of spatial quantum actions or precise orientations of spin. These bound-particle interactions are another consequence of the quantum structure of the inertial system that explains how groups of bound particles with an unchanging geometrical structure can move around independently and act on the natural world as a whole. This regularity holds for complex atoms and bodies composed of them, such as the macromolecules found in living organisms, and thus, the particles of which they are composed have definite positions relative to one another. Quantum physics cannot explain these geometrical structures because it assumes that particles carrying electric charges interact with one another according to basic quantum laws, and the interactions of charged particles are far too complex to derive predictions about their locations from laws of quantum physics. To be sure, quantum physics can make reliable predictions about them, including interactions of macromolecules. But that is because they have a blurry image of bound-particle interactions, called *Density Functional Theories.* These highly mathematical theories abstract from all the detailed interactions of charged particles and use the probability densities described by Schrödinger wavefunctions to describe their structures.[11]

Recognizing how the geometrical structures of composite bodies are constituted by cycles of bound-particle interactions, ontologists will be able to explain the puzzling particle-like nature of electromagnetic waves that led physicists to believe that light is made up of *photons.* Light that interacts with composite bodies appears to be made up of photons because interactions of bound particles in atoms are cycles of precise changes in their relative positions during precise intervals of time that depend on the sizes of temporal and spatial quantum actions in the quantum structure of the mechanical system. Only certain changes in cycles of interactions of bound particles are possible, and when bits of magnetic field matter are absorbed or emitted by atoms, they have waves of certain lengths. Hence, light seems to be particles, called *photons.*[12]

Physicists assume that quantum physics can explain how macro-level

---

[11] There is direct experimental evidence of cycles of bound-particle interactions. There are systems of particles whose lowest energy state is one that switches between detectably different states of this kind in periods of time. These so-called *time crystals* are analogous to unchanging crystals in which particles have periodic relations in space.

[12] Since atoms are the parts of composite bodies, the inertial system explains the photoelectric effect and shows how blackbody radiation depends on Planck's constant. A similar interaction of very short wavelengths of magnetic field matter particles explains the Compton effect. See *Volume I.*

composite bodies, such as levers and keys, can be used as tools because the strengths of the forces between their parts enables them to explain when structures of composite bodies break down. But beneath the probabilistic mathematical surface of quantum physics, there is an expression of the holistic power of matter that does much more, and ontologists will use it to solve the puzzle about the famous double-slit experiment. When particles are shot toward a wall with two closely-spaced slits, the landings of those that pass through the wall produce an interference pattern on a second wall with a wave-like structure, as if the particles passed through both slits like a light wave. This interference pattern is produced even when particles are shot one-by-one, and since no particle can pass through both slits at once, this seems to confirm the wave-like nature of particles described by the Schrödinger wavefunction. But when the reduction of quantum physics to spatio-materialism reveals that subatomic particles go through cycles in which they have precise positions and momentums, it will confirm that the wall is a geometrical cause that constrains what happens by physical causes when it interacts with particles. As the entire wall interacts with a tiny particle, the energy and momentum of both change, and the direction of the particle coming out of the interaction depends on the geometrical structure of the entire wall. Thus, when the wall is rigged to detect which slit the particle passes through, the particle interacts with the detector in the wall, and not surprisingly, no interference pattern is produced.

Though these regularities show how the holistic power of matter solves puzzles in quantum physics, its most far-reaching consequence is how it solves the puzzles in specialized sciences. The inertial system constitutes the motion and interaction of both free and bound particles, and these regularities are different only because the initial and boundary conditions of its quantum structure are different. *Volume II* shows how the expression of the holistic power of matter as geometrical causes solves all the puzzles of sciences studying less-basic regularities about change that have been found in the study of nature. These efficient causes are fittingly called *geometrical* because bodies composed of different kinds of bound particles with the same kind of geometrical structure can constrain physical causes in similar ways. Since these regularities depend more on the geometrical structures than on the kinds of particles of which they are made, overcoming the blindness of physics to geometrical causes will solve all the puzzles of the specialized sciences.[13] This reduction of science to ontology is represented in Figure 6

[13] Geometrical causes are what many scientists now call *information*. Information

by the arrow from *Ontology* to *Science.*

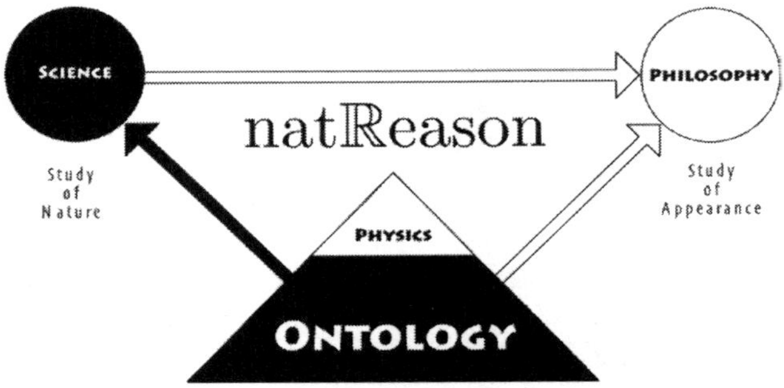

FIGURE 6 - UNIFICATION OF SCIENCE

(2) Geometrical Action

Since geometrical and physical causes are different kinds of efficient causes, they affect what happens in different ways and regularities depend on how they work together. Solutions to the puzzles of the life sciences depend on using geometrical causes to reduce thermodynamics to interactions of space and matter. But steam engines are the simplest example. Heat is the kinetic energy of the motion and interaction of particles on the micro-level, and since steam engines are geometrical causes, they can channel heat to do work on the macro-level. The *second law of thermodynamics* describes a regularity about how much work a given quantity of heat can do on its environment. It depends on a puzzling quantitative property called *entropy*, and the reduction of thermodynamics to ontology will clear up confusion about its nature.

According to the second law, entropy can increase in a closed system, but it can never decrease. Ontology explains increases in entropy as the collective effect of physical causes. Interactions of free particles disperse not only the particles in space but also the (kinetic) matter that moves their rest matter across space, and both are dispersed as evenly as possible among all the particles in the closed system. These ontological causes will reveal that only the dispersal of particles in space (involving volumetric changes) can do thermodynamic work, and the way it is different from the dispersal of

---

is supposed to have a nonphysical, top-down effect, but what happens is actually a bottom-up effect of the interaction of space and matter by which matter has a holistic power. See Paul Davies (2019) *The Demon in the Machine*, Allen Lane, London.

their kinetic matter in time[14] is what explains the puzzling limit on how much work a heat engine can do. This ontological explanation of the thermodynamic regularity studied by physics will clear up a widespread confusion about the nature and role of entropy change in chemistry.

The increase of entropy is the collective effect of physical causes, and since it is a flow of acceleration fuel (the field matter by which particles change one another's motion), geometrical causes can use the flow to do thermodynamic work. In both heat engines and chemical reactions, the geometrical cause must do two kinds of work. One is the kind that acts *on its surroundings*, and the other kind is work that it must do *on itself* in order to do any work on its surroundings at all. Physicists focus on maximizing the work that heat engines can do on their surroundings (which depends on how much work they must do on themselves at a lower temperature). But molecules are micro-level geometrical causes, and since chemists want to know whether chemical reactions will occur spontaneously, they focus on *how much* work molecules must do on themselves (rearranging atoms) in order to do any work *at all* on their surroundings. Ontological scientists will discover that the work that molecules must do on themselves to do any work on their surroundings is an internal volumetric change that evolves into life.

All the life sciences study thermodynamic work. But it can also be called *geometrical action* because it is what happens when geometrical causes constrain the collective effects of physical causes. Since geometrical causes express the holistic power of matter, the particles constrained by geometrical causes are also all related by the puzzling quantum property called *entanglement*. It is not relevant in heat engines since they do work on the macro-level. But the geometrical causes that solve the puzzles of life sciences do chemical work on the micro-level, and what is more, the entanglement of particles in geometrical action on the micro-level is essential to cutting the Gordian knot of philosophical puzzles, as explained in *Volume III*.

In chemical reactions, the flow of acceleration fuel is the collective effect of physical causes, and molecules are the geometrical causes that impose constraints on its flow. Chemical reactions that occur spontaneously do thermodynamic work on their surroundings, and chemists describe them as

---

[14] Since kinetic matter is dispersed *in time* rather than space, changes in it are measured as temperature. Quantities of kinetic matter are the number of temporal quantum actions per unit time, and since it causes a standard distribution of their sizes in time among all the particles in each region of space, it is measured as temperature.

having *Gibbs free energy*. Energy is free because it is possible for acceleration fuel to flow and do work, and the entropy change that chemists must calculate in order to determine whether they have Gibbs free energy is how much work the molecules must do on themselves, rearranging bonds of atoms, in order to do any work at all on their surroundings. The work that molecules must do on themselves is usually an increase in entropy, which I call *Gibbs bound energy*.[15] It is measurable, and so ontological science will discover that the geometrical action that evolves into life is a flow of acceleration fuel doing work in evolving geometrical causes whose quantity can, in principle, be measured as bound energy.

What evolves are geometrical causes of a unique kind, beginning with naturally occurring molecules. They have two basically different kinds of geometrical action, so they are called *double-action geometrical causes*. One geometrical action is the construction of another composite body like themselves, and the other is the construction of a second kind of composite body that can promote the conditions under which they are able to reproduce. When their surroundings drive these geometrical causes through cycles in which they act on the world as a whole in both ways, they acquire powers of doing work in the second way that enhance their power to do work in the first way. This depends on them being located in a region with a finite flow of acceleration fuel that they can use to do thermodynamic work. Since the first kind of geometrical action is reproduction, it increases the population of double-action geometrical causes in the region, and since the accessible flows of acceleration fuel in the region become scarce, some reproductive cycles must eventually come to an end. Those whose secondary way of doing work enhances their power to reproduce are more likely to continue, so double-action geometrical causes change gradually in the direction of greater power to control conditions that make it possible for them to reproduce. If it continues long enough for them to acquire the power to go through such cycles on their own, a form of life begins.

This combination of geometrical and physical causes is clearly a form of natural selection. But I call it the *reproductive mechanism* because population increase is what makes natural selection inevitable in regions with a finite flow of acceleration fuel. It is a reduction of the cause of evolution that

---

[15] In physics, it is the free energy of heat engines, called *Helmholtz free energy*, that determines how much macro-level work they can do, so *Helmholtz bound energy* would be the work that must be done at a lower temperature (returning energy to the surroundings) in order for the heat engine to do any work at the higher temperature.

Darwin discovered to ontological causes because interactions of space and matter explain this combination of efficient causes, and as a more complete explanation of evolutionary change, the reproductive mechanism will enable ontologists to explain the origin of life. That will fill what is widely considered the biggest explanatory gap in physical science except, possibly, for the origin of the Big Bang and the nature of mind.

(3) Origin and Nature of Life

The reproductive mechanism implies that double-action geometrical causes change in the direction of greater functional power *whenever they go through reproductive cycles in a region of finite resources.* This is progressive evolution, and such a region is the surface of a planet like Earth, where the radiation of a star fuels collective effects of physical causes that micro-level double-action geometrical causes can use to act on the world in both essential ways. *Volume II* predicts that ontological science will discover that RNA molecules have powers by which they can do both kinds of work and that they can be driven by the cycle of night and day through endless cycles in which they do work of both kinds. RNA powers are very weak at first. But natural selection gradually makes them stronger, and when *colonies* of RNA are also driven through reproductive cycles by surroundings, such as storms, the functional powers of RNA (and other molecules) also increase by group-level natural selection. Since two levels of double-action geometrical causes going through reproductive cycles are involved in causing their gradual evolution, let us call it a *bi-level embodiment* of the reproductive mechanism. Its recognition will enable ontological scientists to show that there are several *major stages of evolution* on Earth. The first form of life evolves from nonlife during the first major stage, and at three additional major stages of evolution, other forms of life evolve from forms of life that have already evolved. (See Figure 7.)

During the first major stage of evolution, colonies of RNA molecules are driven through reproductive cycles, and ontological science will discover that life begins as prokaryotic cells, such as bacteria. During the three other major evolutionary stages, forms of life that have already evolved give rise to new forms of life. A distinctive development marks the beginning of a form of life during each major stage, and since there are four major stages in the series leading to language-using animals, ontological scientists will see the overall course of evolution as the construction of a four-story *edifice of life*. Beings like us live on the fourth floor.

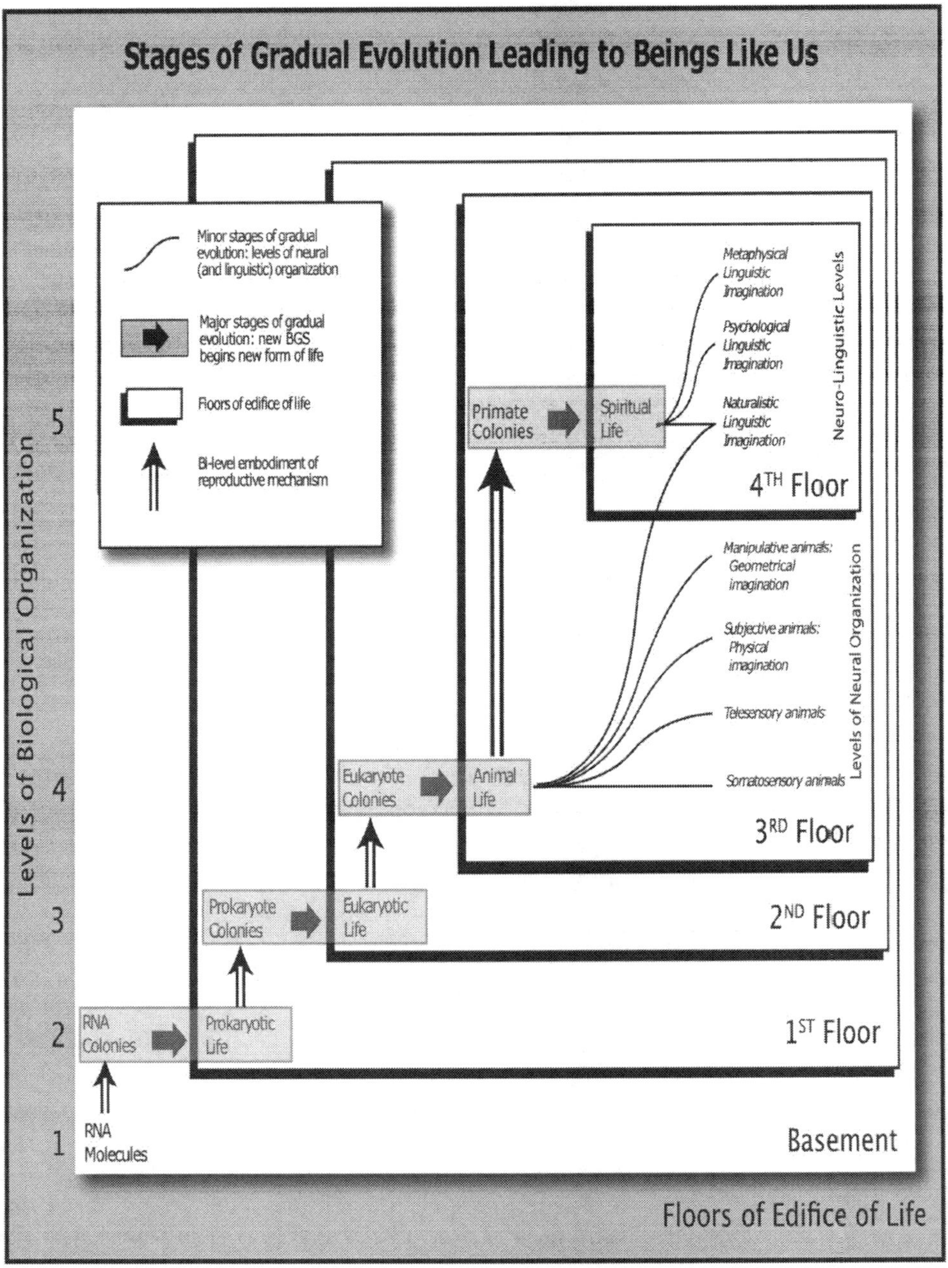

FIGURE 7 - STAGES OF EVOLUTION

The distinctive development that marks the beginning of each form of life reveals the essential nature of life. During each major stage, there are double-action geometrical causes going through reproductive cycles on one level of geometrical organization, and since something drives colonies of them through reproductive cycles, they are double-action geometrical causes on the next higher level of organization evolving by group-level natural selection. Colonies become double-action geometrical causes with ways of doing both kinds of thermodynamic work, and though the work is increasingly complex, they eventually evolve the power to go through reproductive cycles *on their own.* That is, the higher-level double-action geometrical cause eventually acquires a geometrical cause that constrains collective effects of physical causes completely enough to go through an entire cycle of reproduction. That is when a new form of life comes into existence.

Another development that distinguishes the beginning of a form of life from other evolutionary changes is that life causes itself to evolve. Since the ultimate cause of natural selection is scarcity due to population growth, life existing in a region with finite resources embodies the reproductive mechanism. This replacement of the bi-level embodiment of the reproductive mechanism is a radical change in the cause of evolution.

But the evolution of life depends on its double-action geometrical cause going through cycles in which it not only reproduces but also controls conditions that promote reproduction, and since they are completely different kinds of work, the newly acquired geometrical cause must choose between them. To serve that function, the geometrical cause needs three subsystems, each serving a distinct function. One subsystem must register input so that its behavior can depend on the situation. Another subsystem must use input as feedback to carry out each of the various kinds of goal-directed behavior involved in growth and reproduction. And a third subsystem must use input to choose between growing and reproducing. The new geometrical cause with the power to choose between those two incompatible goals can easily acquire the power to choose between other goals involved in growth and reproduction, and since that is what determines its behavior in the situations that arise during each reproductive cycle, I call it the *behavior guidance system (BGS).* Since life begins when a double-action geometrical cause acquires a BGS, the kind of BGS that evolves on each floor of the edifice of life defines the essential nature of that form of life. It is a basic natural kind in the spatio-material world, each with many species.

The BGS is a product of the reproductive mechanism, and when it evolves during a stage of evolution, a double-action geometrical cause turns into a triple-action geometrical cause. The essential structure of a BGS is represented by Figure 8. Notice that a BGS cannot be reduced to a cybernetic system. Though a cybernetic system uses feedback to direct behavior at goals, its goals are given, and so it has no subsystem for choosing between goals. Life begins with the power to choose between two basic goals, and the BGS becomes more powerful by choosing between other goals during each reproductive cycle, each of which may use feedback to direct behavior at the target. But the power to *choose* between goals that help complete a reproductive cycle is the essence of life. And since the BGS is a *choosing machine*, it has *intelligence*, as suggested by the etymology of that term (*inter* means between and *legō* means picking).

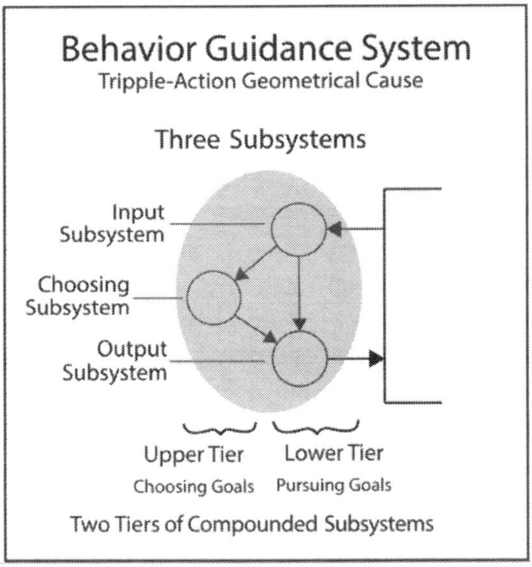

FIGURE 8 - BEHAVIOR GUIDANCE SYSTEM (BGS)

Since the function of the BGS is choosing goals, ontologists will see this geometrical cause as solving the philosophical problem about the compatibility of free will and determinism. In reflective subjects, choices are called *will* because they are represented as intentions, and in the physical world, free will is incompatible with determinism because all events are caused by particles moving and interacting according to the laws of physics.

Assuming that physical laws completely determine what happens, a mathematical derivation from them and the state of the particles at any earlier time can, in principle, predict what will happen at any time in the future, and since a living organism's choice depends on the precise properties of particles in the past, it could not have chosen otherwise.[16] What happens in the spatio-material world is also completely determined. But when living organisms make a choice, they can always have chosen otherwise because choices depend on a geometrical cause whose function is choosing between goals. Figure 9 shows how the unchanging geometrical structure found in a BGS constrains what happens by physical causes so that its behavior is a result of choosing a goal from a range of possible goals. Beneath the probabilistic blurriness of quantum laws, when a geometrical cause interacts as a whole, its interactions with particles change their momentum in ways that deviate from what laws of physics would predict. All particles interact at atomic events that conserve both energy and momentum, and since cycles of bound-particle interactions have atomic events whose relative locations in space and time are fixed, the unchanging geometrical structures of composite bodies help determine the changes in momentum that occur. These deviations from what physical forces otherwise do may be small, but when geometrical causes constrain many physical causes, their effects add up. The BGS guides behavior by channeling interactions of particles in this way. It chooses a goal from a range of possible goals, and though it depends on outside conditions registered by the input subsystem and priorities built into the choosing subsystem, it could always have chosen otherwise because choosing is its function. The precise physical properties of particles extending indefinitely in the past are not causally relevant because a deterministic ontological mechanism that is filtered out of images in the quantum mathoscope constrains physical causes in a way that enables an unchanging geometrical structure of composite bodies to choose goals in a way that serves the function of guiding behavior. Hence, the belief that free will and determinism are incompatible is just an appearance that is caused in physical science by its failure to recognize that, in addition to physical causes, there are geometrical causes constraining what physical causes do.

---

[16] Since quantum laws are probabilistic, such derivations are not possible, and though that does not entail indeterminism, even if quantum regularities were indeterministic, that would not explain how choices are free because they would still depend on past properties of particles that are irrelevant to the choice.

Life begins with the evolution of a BGS, and that is a radical change in the course of evolution because life causes itself to evolve in regions with finite resources. What causes a form of life to evolve is a bi-level embodiment of the reproductive mechanism. It causes colonies of double-action geometrical causes to evolve gradually in the direction of greater functional power; the colonies become double-action geometrical causes; and when they acquire a BGS, life itself becomes the ultimate cause of its gradual evolution in the direction of greater functional power in a region with finite resources. By going go through reproductive cycles on their own, they increase their numbers, and since their population growth eventually requires some reproductive cycles to come to an end, they impose natural selection on themselves. And since the BGS at the core of life is a choosing machine, the evolution of life brings a kind of property into existence whose nature has long been puzzling to beings like us.

The origin of life reveals the essential nature of life, and as ontological scientists will recognize, the nature of life, in turn, explains the nature of the good. To be alive is to be a BGS choosing goals, and since choosing goals is what matters to a living organism, the goals that their BGS normally chooses are what all living organisms would call good, if they could speak. Since BGSs evolve by the reproductive mechanism, good goals tend to promote conditions under which they can reproduce, and the goals they normally choose change in the direction of maximum power to control relevant conditions. Hence, goals necessarily contribute to functional power. But that is not what makes them good. That is just what makes them a goal that a BGS can choose. What makes a goal good is that it is the one normally chosen from among other possible goals by the BGS. Hence, goodness is entailed by the nature of life. Life begins with the evolution of a choosing machine, and from its point of view, goals are necessarily good or bad. Choosing goals that are good is what matters to living organisms. Indeed, life can be defined as *geometrical action aimed at the good.*

Evolutionary biology is blind to the nature of life and the nature of the good because physical science is blind to geometrical causes. Since the possibility of double-action geometrical causes is hidden from Darwinists, they must assume the existence of organisms that reproduce themselves to explain evolutionary change, and so the origin of life from nonlife is a great puzzle. But knowing that double-action geometrical causes are what evolves, ontologists will explain evolution from the beginning as a change that gives them every possible functional power as it becomes possible for random variations to try it out. Since gradual evolution is change in the direction of greater functional power, it is progressive. And ontologists will recognize that evolution is progressive in another way that is also conceivable in Darwinian biology. Since what evolves are geometrical causes, levels of geometrical organization can cause stages of gradual evolution with step-like increases in functional powers.

(4) Stages of Evolution

Understanding how the reproductive mechanism causes the origin and nature of life, ontological scientists will know that it can also cause stages of evolution. The reproductive mechanism includes all the conditions that cause double-action geometrical causes to evolve in the direction of greater functional power, and it implies that another stage is inevitable whenever a higher level of geometrical organization in evolving geometrical causes is both possible and functional. It is *possible* when it can be tried out as a random variation on evolving geometrical causes, and if it is *functional* in the sense of opening up a range of powers for controlling conditions affecting reproduction that were previously out of reach, double-action geometrical causes on the higher level will eventually begin to evolve gradually in the direction of greater functional power. They will acquire every possible functional power as it becomes possible during gradual evolution. Using the origin of life as a model for explaining stages of gradual evolution, ontological scientists will know that they depend on a bi-level embodiment of the reproductive mechanism, and their most surprising discovery may be that a series of inevitable stages brings beings like us into existence. A bi-level embodiment of the reproductive mechanism causes a new form of life to evolve on the upper level of reproductive cycles of double-action geometrical causes, and since the essential nature of a form of life is defined by its kind of BGS, this series implies that language-using animals like us are parts of a previously unrecognized form of life that evolves on the fourth floor of the edifice of life.

During the first major stage of evolution, RNA are driven through reproductive cycles on the lower level by the cycle of night and day, and since there are colonies of RNA whose surroundings somehow drive them through cycles of reproduction, the colonies agglutinate a BGS on the upper level. It is little more than a cell enclosing a loop of DNA together with all the kinds of molecules needed for molecular interactions to serve the three subfunctions of guiding behavior: input, choosing, and output. They are *prokaryotic cells*. (See Figure 7.)

During the second major stage of evolution, cycles of reproduction are somehow imposed on colonies of prokaryotes that are going though lower-level reproductive cycles on their own, and those colonies likewise agglutinate a BGS on the upper level. It is also a cell, but it is far more complex. A DNA molecule determining all the kinds of thermodynamic work that molecules do is contained in a nucleus inside the cell, and the way that it gradually evolves during the stage gives it the power to reproduce the cell not only asexually (by simply duplicating its own DNA) but also sexually (by mixing segments of DNA derived from two different cells). They are *eukaryotic cells*, and since the mixing of DNA increases the range of possible random variations that are tried out, sexual reproduction expands the range of powers that can evolve and accelerates the rate at which functional powers are acquired.

Organisms composed of many eukaryotic cells are the form of life that evolves during the third major evolutionary stage. It is basically different from life on the first two floors of the edifice of life because instead of using a *cell* as the geometrical cause that enables interactions of other parts to serve as a BGS, it uses a *plan-sharing geometrical cause*. When eukaryotic cells go through reproductive cycles in a colony and their offspring remain in contact, their interactions can generate colony-level behavior. When some condition of their environment drives colonies of eukaryotes through reproductive cycles, genetic plans evolve that give colony-level behavior functional powers. When the plans enable the asexual reproduction of a single cell that has reproduced sexually to populate colony, its daughter cells all have the same genetic plan, and a new form of life begins when the colony-level behavior includes sexual reproduction with cells from another colony because they go through reproductive cycles on their own. Since the genetic plan is shared by the asexual reproduction of a mother cell, its sexual reproduction plays the same role in the plan-sharing BGS of the third-floor form of life as the cell plays in the cell-sharing BGSs of forms of life that evolve on the first two floors of the edifice of life. This is the origin of plants, fungi, and animals, depending on the kind of flow of acceleration fuel that the plan-sharing BGS uses to do thermodynamic work. And since animals are heterotrophs, the genetic plan must construct a body that can act on other objects in space, and thus, multicellular animals have an essential nature that makes it possible for another kind of plan-sharing BGS to bring a new form of life into existence on the fourth floor of the edifice of life.

To go through cycles of sexual reproduction, multicellular animals must mate as well as acquire acceleration fuel from some source in the region, and to guide the behavior of an animal body through cycles in which it does both essential kinds of work, the animal needs a BGS built into the animal body. Thus, the *animal* (plan-sharing) *BGS* builds a *neural BGS* into the body, and the structure of the nervous system is mainly responsible for the range of functional powers that animals can acquire. They become very powerful because, as we shall see, a series of levels of neural organization can cause a series of minor stages of evolution at each of which a new range of functional powers evolves in animals. Some kinds of animals live together in groups, and they are the colonies of animals going through reproductive cycles on the lower level that enables a bi-level embodiment of the reproductive mechanism to bring a new form of life into existence on the fourth floor. Reproductive cycles can be imposed on groups of animals by their population growth because the scarcity it causes can require them to divide into smaller groups. When the animals are powerful enough, they can acquire the use of language, and since that is a way of sharing a plan, a BGS based on sharing a cultural plan can coordinate their behavior in a way that enables groups of them to go through reproductive cycles on their own. Thus, groups of animals become the new form of life on the fourth floor of the edifice of life. That is the origin of human societies. Since they are a product of a bi-level embodiment of the reproductive mechanism, they are a form of life in the same sense as prokaryotes, eukaryotes, and multicellular organisms. But since the language-using animal bodies of which they are composed is the only body of this form of life, it is fitting to call them *spiritual organisms.* Furthermore, it is a uniquely powerful form of life because the spiritual BGS is itself an embodiment of the reproductive mechanism. The behavior of spiritual organisms is coordinated by linguistic interactions, and as we shall see, since the linguistic representations exchanged are double-action geometrical causes going through reproductive cycles in a region where some cycles must come to an end, the evolution of the cultural plan that coordinates the members' behavior is the main cause of spiritual organisms acquiring new powers that enable them to control conditions that affect their reproduction.

The edifice of life is just the gross structure of the stages of evolution that lead to beings like us. Ontological scientists will discover that during the major stages that bring new forms of life into existence on the third and fourth floors of the edifice of life, there are minor stages of evolution caused by levels of geometrical organization in the nervous systems. The animal BGS builds a nervous system into the animal body for guiding its behavior, and since the neural BGS is a geometrical cause that helps the animal BGS serve its function of choosing between goals in a way that enables them to go through reproductive cycles on their own, the neural BGS also has three subsystems serving each essential subfunction (input, choosing, output). When a higher level of organization in evolving geometrical causes is both possible and functional, another stage of evolution is inevitable, and ontological scientists will discover that a series of levels of organization in neural BGSs causes a series of minor stages of gradual evolution on the third and fourth floors of the edifice of life that lead inevitably to the existence of beings like us.

On the third floor, there are four neural levels and four minor stages. On the first neural level are somatosensory animals. In them contact with other objects enables behavior to control relevant conditions, and so *single neurons* can serve each of the three essential behavior guiding subfunctions. Telesensory animals, on the second neural level, have *bundles of neurons* serving each subfunction, and with sensory organs that can detect objects at a distance from their bodies, a centralized nervous system enables them to guide locomotion and other kinds of behavior to goals that are good. But when each subfunction is served by a *circuit of bundles of neurons* in the brain, the subsystems can inter-act with one another, and that enables the third level of neural organization to serve as a *faculty of imagination.*

Knowing that levels of neural organization can cause minor stages of animal evolution, ontological scientists will trace the structure of the mammalian forebrain to anatomically distinct homologous structures in the nonmammalian vertebrate brain and discover that the basic structure of the mammalian brain has the function of serving as a faculty of imagination. In the nonmammalian brain, it is obvious that the midbrain, forebrain, and hindbrain serve, respectively, the input, choice, and output subfunctions of the neural BGS, and evolutionary changes in the vertebrate brain show that, in mammals, their functions are taken over by three circuits in the forebrain: the sensory input circuit, the choosing circuit, and the motor output circuit.[17]

---

[17] In telesensory vertebrates, called *mammal-like reptiles,* the midbrain, forebrain, and hindbrain are the subsystems serving each of the essential subfunctions, and mammals evolve from them by relocating each of the subsystems (and their connections to one another) to the forebrain where each becomes one of the three circuits. Each circuit is many bundles of neurons that all pass through at least three major structures of the mammalian forebrain: a region of its distinctive surface (called the *neocortex*); the thalamus at the center of the forebrain (where the spinal cord and brain stem end), and a third distinctive structure in the forebrain whose connections show their homology to structures of the nonmammalian vertebrate brain. The *sensory input circuit* is located

The mammalian forebrain can serve as a faculty of imagination because the motor output circuit can generate behavior covertly as well as overtly. When behavior is only covert, it calls up memory images recorded by the sensory input circuit, anticipating consequences of the behavior. The choosing circuit determines goals of behavior by attaching desires to objects represented in those images, and since the mammalian brain can anticipate consequences of choosing goals, it can plan behavior prior to overt action. Mammals are properly distinguished from other animals as *subjects* because their faculty of imagination enables them to perceive their bodies as objects with a location in a world of objects in space. The subjective neural BGS enables mammals to guide their locomotion relative to other objects in space, to plan how to move their bodies around in the natural world in ways that will satisfy their desires, and more generally, to understand the effects of motion. That is, it enables them to understand basic *physical causes* in the natural world. See Figure 9.

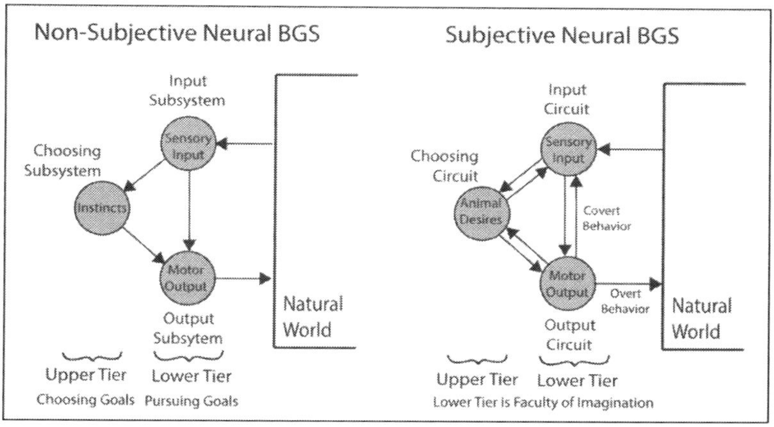

FIGURE 9 - NEURAL BGS WITH FACULTY OF IMAGINATION

---

mainly in the posterior cerebrum (though it also has connections with the other parts of the neocortex by which it records memory images). The *motor output circuit* is located in the anterior cerebrum, and since it can generate both overt or covert behavior, its inter-action with the sensory input circuit is the faculty of imagination that represents consequences of covert behavior. The *choosing circuit* is located in both the posterior and anterior cerebrum, so that it can use sensory input to determine goals for behavior and the plans for attaining them. For details, see *Volume II, Chapter 7*.

Physical causes are only one kind of efficient cause at work in the natural world, and so there is another entire range of functional powers that a faculty of imagination can give mammals. Ontological scientists will discover that the primate neural BGS has several circuits within each of the three circuits of bundles of neurons in the mammalian forebrain. Each sub-circuit guides the behavior of one of four hands, and since that enables primates to manipulate objects in space, they have a faculty of imagination that enables them to anticipate how shapes of composite bodies will affect their interaction. That is, it enables them to understand basic *geometrical causes*. This is the fourth level of geometrical organization in neural BGSs, and so these levels determine four different natural kinds of animals.[18]

(5) Spiritual Life

There is a fourth floor to the edifice of life because a yet higher level of neural organization enables groups of social primates to acquire a plan-sharing BGS for guiding their behavior based on the use of language. Groups of social animals are already double-action geometrical causes because they can reproduce by division and they jointly promote conditions, such as the growth of their populations, which makes it possible for groups to reproduce. When they acquire the use of language to coordinate their behavior, they become a form of life, called *spiritual organisms*, because members use language to share plans by which they go through cycles in which they do both essential kinds of thermodynamic work. Like all forms of life, spiritual organisms cause their own evolution in regions of finite resources. Ontological science will discover that the first two minor stages of spiritual evolution depend on the evolution of a higher level of organization in the neural BGS, which is manifested as a level of linguistic organization. The first is the use of *naturalistic sentences*, whose simple subject-predicate (or subject-predicate-object) grammar can represent states of objects in space and describe their causal relations, for example, *Deer are grazing on the other side of the hill*. The second is the use of *psychological sentences*, which predicate psychological states of subjective animals and can describe how their beliefs and desires cause their behavior, for example, *He believes that deer are grazing on the other side of the hill, but he is mistaken*.

---

[18] Bilateral animals fall into to two basic groups depending on their kind of embryological development. Vertebrates are *deuterostomes*, and invertebrates are *protostomes*. Since protostomes can evolve only the first two levels of neural organization, invertebrates never acquire a faculty of imagination like mammals.

The higher level of neural organization on which the use of naturalistic sentences depends is constituted by an interaction between the two hemispheres of the mammalian cerebrum. The left hemisphere usually handles the sentences themselves, leaving the constructions of their meanings up to the right (nonverbal) hemisphere. The use of psychological sentences depends on a yet higher level of neural organization. It is a geometrical cause built into the frontal lobe of the verbal (left) hemisphere that enables speakers to temporarily impose the meanings of sentences on their neural BGSs and use their meanings to think about how psychological states, such as beliefs (input), desires (choice), and intentions (output), affect one another. Since speakers can tell when they already have the psychological states that they impose transiently, they can reflect on the causes of their own choices in the process of making them, and since they can also predicate psychological states of other subjects, they can simulate the reasoning of others. Thus, they can see into the minds of others by inferring their psychological states as the best explanation of their behavior (which is much easier when others can speak).

Distinguishing these linguistic levels will enable ontologists to explain the difference between archaic and modern humans. Archaic humans (such as Neanderthals) were members of spiritual organisms at the *naturalistic level*, and they could follow a leader. Members of spiritual organisms have a desire, derived from the dominance hierarchy in social animals, to do what is prescribed by authoritative linguistic representations, and thus, the use of naturalistic sentences enabled a leader to assign tasks to members that coordinates their behavior in pursuit of goals that are good for the spiritual organism, such as preying on large animals and fighting wars with other naturalistic spiritual organisms. Though archaic humans had functional powers far beyond those of groups of primates without the use of language, they became extinct because they tapped the same flows of acceleration fuel through the surface of the planet as modern humans. The power of groups to hunt large animals meant that competition for scarce resources led to war, and the higher order of functional powers that evolved in modern humans enabled them to replace archaic humans in every ecological niche. Subjective animals with the use of psychological sentences are reflective subjects who can explain the behavior of other subjects by the psychological states that cause it, and since they could *see into the minds* of members of archaic humans, modern humans caused the extinction of naturalistic spiritual organisms. Moreover, since reflective subjects can think about the goals of behavior, modern humans knew that they shared certain goals with other members of their groups, and so instead of simply following the orders of a leader, they could cooperate in the pursuit of goals that are good for the group as a whole. By arguing about how to coordinate behavior in pursuit of shared goals, they were better able to discover good plans, and they were more powerful because they understood their plans in a way that allowed them to improvise in unexpected situations.

During minor stages of spiritual evolution, gradual evolutionary change is more rapid because the exchange of linguistic representations that guides the behavior of spiritual organisms is itself an embodiment of the reproductive mechanism. Linguistic representations can be double-action geometrical causes because when they are understood, they are reproduced in other brains, and when they are accepted, they can determine how members behave. Since the number of other brains is finite, linguistic representations go through reproductive cycles in a region that requires some reproductive cycles to come to an end, and their general acceptance can guide the behavior of organisms with a plan-sharing BGS based on language. Since agreement depends on the members' judgments, culture evolves by *judgmental selection.* At the first stage of spiritual evolution, naturalistic sentences are used mainly to agree about means to ends that are shared but not represented as goals that are chosen. Thus, cultural evolution is mainly technological, for example, about how to use tools, how to hunt, and which kinds of food to gather. But at the psychological level, culture evolves further and more quickly because reflective subjects who can think about alternative goals can argue about which goals to pursue, not just how to pursue them. Thus, while naturalistic level culture evolves by *technical selection*, psychological-level culture evolves by *pragmatic selection.*

Since the use of a language with psychological sentences enables members to reflect on the psychological states that cause their beliefs and behavior, reflective subjects can tell stories about who they are, where they come from, and what they intend to do. Hence, they have a *self* with interests that help them choose goals that are good for them. Since they can see into one another's minds, they recognize that others also have a self and interests that are good for them, and rules evolve for governing their interactions that promote conditions under which they are motivated to cooperate with one another. They are *moral rules*, and since the essential nature of life explains the nature of the good, the essential nature of the spiritual form of life explains why reflective subjects ought to be moral. Choosing to do what is right rather than what is wrong is what matters to reflective subjects because that is the choice they must make in order to lead a life as part of a spiritual organism.

Moral rules accumulate along with other rules as part of culture, and ontological scientists will explain the ability of reflective subjects to follow them by a desire to conform to linguistic input with spiritual authority that derives from the dominance hierarchy of social animals. When a public competition determines the leader of an animal pack, the loser acquires a desire to obey the victor, and on this model, the spiritual authority of a leader's commands depends on a *spiritual desire* that motivates members of naturalistic spiritual organisms to obey their leader. But in psychological spiritual organisms, shared rules are authoritative spiritual input to members, and since reflective subjects can represent the psychological states causing their behavior as part of the process of choosing it, they can represent what they choose as their will. And the will is free because, like all forms of life, they could always have chosen otherwise. Choosing between goals is the function of the BGS, and since it is a geometrical cause that constrains physical causes, choices cannot be predicted by derivations from laws of physics and precise physical properties of particles in the past. But since reflective subjects have a will, they know when they are acting on their spiritual desire and following rules and when they are not, and so they can take responsibility for their choices. When they do wrong, they know that they are responsible for what they do, and when others hold them responsible, they learn to use their spiritual desire to follow rules. They can choose to follow authoritative spiritual input in the future, and using the model of the dominance hierarchy in packs of animals, ontological scientists will recognize that punishment strengthens their spiritual desire. When the leader of the pack defeats an alpha male's challenge, his rebellious desire is weakened and his desire to submit to authority is strengthened. This brain mechanism is recapitulated when parents punish their children for disobedience. Since they understand how they are responsible for what they do, they learn to act on their spiritual desire and take responsibility for what they choose. The authority of the rules is internalized as part of the conception that reflective subjects have of the kind of self they aspire to be, and thus, when they do what is morally wrong or fail to follow other parts of the cultural plan that they have internalized, their spiritual desire causes them to feel guilt or shame.

(6) Civilization

The reproductive mechanism implies that the essential natures of the forms of life depicted in Figure 7 are all determined by the levels of geometrical organization that cause the evolutionary stages during which they evolve. Each stage is a progressive gradual evolution, and the series of stages is a second way that evolution is progressive because at each stage, living organisms acquire an entire new range of functional powers. During the psychological stage, the last stage explained in the ontological reduction of science, evolution is mostly a gradual change in culture caused by pragmatic judgmental selection. But during that minor stage, there is another comparably significant inevitable development: civilization. Though it is a cultural development, it has another cause.

Civilized spiritual organisms depend on the evolution of a culture whose rules give members of two groups determined by birth different kinds of rights relative to pieces of land throughout a territory. This is what Marx called a *class structure*, and it depends on a *forcefield* of spiritual authority linked to the land, such as a feudal system. There is often another group of reflective subjects, called *slaves.* They are not members of spiritual organisms because members treat them as property (and since they are taken as prisoners in war, slavery can occur without civilization). But the evolution of civilization is inevitable because, when a class structure is tried out that enables spiritual organisms to coordinate the behavior of huge populations over indefinitely large territories, the sheer number of members opens up an entire range of new functional powers for spiritual organisms to evolve. Ontological scientists will recognize that civilization is analogous to eusocial animals, such as insect colonies, on the third floor, because, on each floor, they are a second level of geometrical action generated by the kind of BGS that evolves on its floor.

Since writing is one of the powers that civilizations typically evolve, historians can describe its rise. But ontological scientists will describe the rise of civilization as inevitable because they will know the essential nature of the object studied by history. It is a spiritual organism with a class structure that enables the evolution of an entire range of new functional powers. Historians also describe the fall of civilizations, and since the causes are contingent conditions, they could have been avoided. But ontological scientists will not find a cause that makes the fall of civilization inevitable. On the contrary, in Western civilization, they will find evidence that a fall is not inevitable. The Enlightenment spawned not only physical science, but also liberal political institutions and an economic system that uses technological discoveries to increase wealth indefinitely over generations. The functional powers of these institutions are so obvious that they are still spreading peacefully to other civilizations on Earth. Furthermore, even ontological science itself will be evidence that Western civilization is not vulnerable to the causes of decline in other civilizations. It derives from physical science, and since it will have explained everything found in the study of nature, it will be evidence that the West's functional power can continue to increase.

Chapter 3

Unification of Science and Philosophy (Volume III)

Theoretical Reason

The reduction of the laws of physics to ontological causes has far-reaching consequences. It reveals a holistic power of matter, and after solving the intractable puzzles of specialized sciences, it solves the intractable puzzles of philosophy. It entails a kind of efficient cause not recognized by physics, and as *Volume II* argues, knowledge of how geometrical and physical efficient causes work together will enable ontological scientists to reduce science to ontology. But as predicted in *Volume III*, the holistic power has a further significance. A combination of the two kinds of efficient causes as the reproductive mechanism is responsible for the evolution of a special kind of geometrical action in the mammalian brain, and assuming that matter has a primitive phenomenal intrinsic property, ontological scientists will explain how consciousness is part of the natural world in a way that unifies science and philosophy.

The reduction of laws of physics to spatio-materialism will reveal that reflective subjects are an inevitable product of the reproductive mechanism and that civilization is an inevitable development during the psychological stage of spiritual evolution. But that will not be a complete explanation of what is found in the natural world because Western civilization has functional powers unlike those of other civilizations. Ontological science itself will be evidence that, far from heading for a fall, it is able to continue evolving in the direction of greater functional powers.

Ontological science is the reduction of physics to ontology, and as described in *Part One*, its discovery depends on correcting a mistaken methodological assumption of physics. Its laws are formulated mathematically, and empirical lawism assumes that mathematics is known independently of perception. But mathematics can, in principle, be falsified by perception. As ontologists will have discovered, mathematics is true by correspondence to a world constituted by space and matter because that is the best explanation of the "unreasonable effectiveness" of mathematics in discovering laws of physics. Knowing that ontological science derives from physical science, ontological scientists will trace their own origin to the Enlightenment and Kant's defense of *a priori* mathematics. They will know from Kant that confidence in Reason is a legacy of two millennia of arguments about metaphysics, and their own explanation of mathematical truth will show that its belief in rational intuition is false. Thus, the challenge will be to explain how a false belief about how mathematics is known could have made physics so successful for centuries—only to lead in the end to intractable puzzles that physics could not solve.

Metaphysics is the belief that Reason can know Reality behind Appearance, and since Western metaphysicians assumed that knowledge depends on faculties of intuition, they were intuitionists. Since ontological scientists are conscious reflective subjects, they will be able to understand intuitionistic metaphysical arguments. They will know that the reflective neural BGS causes everything they believe or say because their ontological explanation of the series of inevitable stages of evolution that brings reflective subjects into existence reveals how mammalian brains work. What intuitionistic metaphysicians believe about the causes of their behavior is false, but since ontological scientists will know that the natural world is constituted by space and matter, they will be able to explain why metaphysicians hold that false belief. They will infer that matter has a phenomenal way of existing in itself as the best explanation of how consciousness is part of the natural world. That will reveal an illusion inherent in consciousness that explains how exchanging metaphysical arguments causes culture to evolve an entire range of functional powers that were previously out of reach. And knowing that physical science was inspired by the confidence in Reason built up from two millennia of philosophers arguing about metaphysics, they will explain Western civilization as a stage of spiritual evolution that follows the stage at which reflective subjects evolve. This ontological explanation of ontological science as an inevitable product of the metaphysical stage will turn ontological science into naturalistic Reason.

Science is the use of the empirical method to *Study Nature*, while philosophy uses another way of knowing to *Study Appearance*. Their unification is represented in Figure 10. The arrow from *Science* to *Philosophy* represents the contribution of the reduction of science to ontology, while the arrow from *Ontology* to *Philosophy* represents a more direct contribution from ontology. Together, they will explain Western philosophy, and the dawning of this new self-understanding of science will be the culmination of gradual evolution during the metaphysical stage, called the *Second Enlightenment*. Ontological scientists will find themselves knowing Reality behind Appearance, and ontological science will become naturalistic Reason.

FIGURE 10 - UNIFICATION OF SCIENCE AND PHILOSOPHY

This prediction is defended in *Volume III*. Once ontological scientists trace the origin of ontological science to Enlightenment confidence in Reason, they will discover that metaphysics was the motor behind Western history. But Reason has a practical function as well as a theoretical function, and the Second Enlightenment will perfect both. Both are represented in offspring of the 18th-Century Enlightenment. Science will evolve into the perfect theoretical knowledge to which philosophy always aspired, and together with the evolution of liberal political institutions and capitalism, naturalistic Reason will turn out to be the perfect way of knowing to which the ancient Greeks aspired. This chapter describes the cultural changes that culminate in naturalistic Reason serving its theoretical function perfectly, and the final chapter describes the changes that enable it to serve its practical function perfectly.

(1) How Consciousness is Part of the Natural World

Consciousness is the *phenomenal* character of experience. When beings like us perceive the natural world, for example, simple sensory *qualia*, such as colors, sounds, odors, tastes, and itches, and other bodily sensations, are immediately present. Qualia are *simple phenomenal properties* because they are kinds of phenomenal properties that can be defined only by pointing to instances of them. But when we perceive the natural world, many qualia of various kinds seem to be located in a 3-D space, so they are complex phenomenal properties. Moreover, since we have bodies located in a world of objects, the way that sensory qualia are configured in space and change as our bodies move around makes them seem to be *qualia bodies* in a world of *qualia objects in phenomenal space*. That is the phenomenal character of perception, and since we are still conscious when we use psychological sentences to reflect on our psychological states, there is a fundamental difference between *reflection* and *consciousness*.

For physical science, the existence of sensory qualia is a hard problem because laws of physics do not mention phenomenal properties. But it is not hard for ontological science. Since it holds that the natural world is constituted by space and matter, it can explain how phenomenal properties are part of the natural world as what it is like to be a substance.

As parts of the natural world, reflective subjects are constituted by substances, so their consciousness can be explained as what it is like to be a substance helping constitute them. This is an ontological explanation of phenomenal properties, and ontological science will discover it by an inference to the best explanation. Since ontological scientists are reflective subjects, they can describe their consciousness, and the best explanation of why they are conscious is that matter has a phenomenal way of existing in itself. Since this phenomenal property must be how all the matter in world exists in itself, they will accept a form of *panpsychism*. Hence, it must be like something to be even the simplest bits of matter. But reflective subjects cannot know what it is like, and since an isolated bit of matter is not a subject, their hypothesis will be that matter has a *proto-phenomenal intrinsic property* of some kind. This is a proper ontological cause because it is possible that the existence of a substance entails the existence of a qualitative property. But if this property is nothing but what it is like to be a bit of matter, it is not an efficient cause. Though it is an intrinsic property of the bit of matter, it is not the basis of a power that is expressed by how bits of matter interact with other substances, and since phenomenal properties make no difference in what happens, they are called *epiphenomenal* properties. *Epiphenomenal panpsychism* solves the hard problem of mind, and by using it to explain why complex phenomenal properties are immediately present to mammals perceiving the natural world, ontological scientists explain how consciousness is part of the natural world.

The arrow from *Ontology* to *Philosophy* in Figure 10 represents the solution to the hard problem of mind by epiphenomenal panpsychism, and as the diagram indicates, it is only part of the explanation of how consciousness helps metaphysics cause a stage of spiritual evolution that follows the psychological stage. The arrow from *Science* to *Philosophy* represents the other part. The ontological explanation of evolution reveals the series of inevitable evolutionary stages that brings reflective subjects into existence, and its explanation of the evolution of the mammalian brain reveals that its structure serves as a faculty of imagination.[19] This faculty is the interaction of a motor output circuit of bundles of neurons in the anterior cerebrum with a similar circuit in the posterior cerebrum that registers sensory input and lays down memory images throughout the cerebrum. The motor output circuit generates overt behavior that causes changes in sensory input. But when it generates behavior covertly, it calls up memory images in the sensory input circuit representing its consequences. With the ability to anticipate what will happen, mammals can plan behavior before acting overtly. Mammals are properly called *subjects* because the registration of sensory input in the faculty of imagination enables them to perceive their bodies as objects located outside other objects in a three dimensional world, and more generally, their ability to predict consequences of motion is an understanding of basic physical causes.

The combination of the arrows from *Ontology* and *Science* to *Philosophy* in Figure 10 represents the ontological explanation of consciousness. Given the ontological cause that explains how phenomenal properties are part of the natural world, ontological scientists will use their explanation of how the mammalian brain works[20] to infer that an intrinsic property of some bit of matter helping constitute the mammalian brain is the best explanation of why mammals are conscious. There is one—and only one—species of matter helping constitute geometrical action in the mammalian brain that can explain consciousness. That is the magnetic field matter constituting what

---

[19] The existence of this faculty of imagination is described in neuroscience as the so-called *default mode network*, a number of distinct areas in the neocortex that MRI scans show are regularly interacting with one another when the brain is not performing a task. It is seen as *mind wandering*. It is not seen as the default activity of the geometrical cause that evolves at the subjective stage of evolution because it is able to serve as a faculty of imagination. Neuroscientists are blind to this cause because they are physicalists who do not recognize geometrical causes, much less stages of evolution.

[20] They can be sure that the basic structure of the mammalian forebrain serves as a faculty of imagination because as *Volume II* shows, at the third stage of animal evolution, three anatomically distinct subsystems of the nonmammalian vertebrate neural BGS that serve the three essential subfunctions of the neural BGS evolve into the three circuits of bundles of neurons embodying the faculty of imagination in the mammalian brain.

physicists call the *electromagnetic field* generated by brain activity. In the spatio-material world, all species of matter are defined by the spatiotemporal structures of their coincidence with space, and magnetic field matter is the species spread out continuously in the electromagnetic part of the inertial system that mediates interactions of charged particles.[21] Since configurations of sensory qualia in phenomenal space are immediately present to mammals perceiving the natural world, consciousness is what it is like to be a mammal.

This discovery about how consciousness is part of the natural world depends on the holistic power of matter in two ways, first, as the geometrical action of the mammalian brain, and second, as a phenomenal intrinsic property of that geometrical action that depends on matter having a proto-phenomenal intrinsic property.

The mammalian brain is a geometrical cause, and its geometrical action is what happens as it constrains collective effects of physical causes everywhere in the brain. Laws of electromagnetism describe the regularity by which the acceleration of charged particles causes waves of field matter that are carried across the inertial system at the speed of light. The firing of a neuron is a rapid acceleration of charged particles (called *ions*) in and out of its outer membrane, and its propagation along the neuron's axon is called an *action potential*. It causes waves of magnetic field matter, and they help determine interactions of ions in neurons all along its path. Furthermore, neuron firings throughout the mammalian brain are synchronized (mainly by the thalamus projecting fibers to both the anterior and posterior neocortex). Thus, magnetic field matter in every small volume of the mammalian forebrain has a spatiotemporal structure that registers the firing of neurons everywhere in the brain, and since its basic structure serves as a faculty of imagination that enables mammals to see their bodies as located in a world of objects in space, all the small volumes of magnetic field matter have a spatiotemporal structure that is isomorphic with the representation of the natural world that it constructs in the sensory input circuit. Though the spatiotemporal structures of field matter in different small volumes of

---

[21] The relevant part of the neural geometrical action includes only matter in the electromagnetic system because bits of matter in the mechanical system, the other part of the inertial system, are not spread out in space. The mechanical system has a quantum structure that helps constitute particles and their motion by dividing matter up into distinct bits whose quantities are measured by their numbers of cycles of quantum actions per unit time.

the brain are slightly different from one another, they are continuously connected at the speed of light, and so ontological scientists will infer that consciousness is what it is like to be the magnetic field matter helping constitute the geometrical action of the mammalian brain.

The geometrical action of the mammalian brain is the first expression of the holistic power of matter because the unchanging structure of the mammalian forebrain is a geometrical cause. Since it is responsible for the neural firings throughout the forebrain, it gives magnetic field matter a spatiotemporal structure that is isomorphic with the complex phenomenal properties that are immediately present. But these phenomenal properties also depend on a further expression of the holistic power of matter. When matter has a proto-phenomenal intrinsic property, the holistic power is also expressed by complex phenomenal properties that are immediately present to mammals perceiving the natural world. This is consciousness, and there is a *unity* to consciousness because many qualia are parts of each phenomenal property.[22]

---

[22] Consciousness is the phenomenal way that matter helping constitute the geometrical action of the mammalian brain exists in itself, and to understand how this geometrical action expresses the holistic power, it helps to distinguish two different aspects of the unity of consciousness. First, the spatiotemporal structure of its coincidence with space explains why qualia are configured in phenomenal space when mammals perceive the natural world. If a body composed of neurons in a similar way did not serve as a faculty of imagination, but had some other kind of geometrical cause, sensory qualia would not be configured in phenomenal space in a way that makes it seem that they are a world immediately present to a subject. Indeed, what it is like to be that body might not involve having qualia at all. This structure of the unity of consciousness is the *geometrical expression* of the holistic power of matter. Second, there is a more basic aspect of the unity of consciousness because many qualia of various kinds are all immediately present at once, and this is the geometrical expression of the holistic power that depends on matter having a proto-phenomenal intrinsic property. The magnetic field matter that helps constitute the geometrical action of the mammalian brain has a complex spatiotemporal structure in every small volume, and since the many sensory qualia of various kinds located in phenomenal space are all immediately present to the mammal at the same time, they give consciousness the *unity* implied by calling them *complex phenomenal properties*. But the ions could be just a plasma (that is, just many charged particles moving independently of one another), and even if they happened to be moving and accelerating in the same way at some moment as ions in the mammalian brain, the magnetic field matter waves they generated would not have the kind of phenomenal intrinsic property that mammals have when they

This ontological explanation of consciousness implies that it has another relevant property. Since the basic structure of the mammalian brain is what enables it to serve as a faculty of imagination, the way that mammals perceive makes it seem that sensory qualia *are* the natural world. That is an illusion because perception actually depends on input from objects in space to sensory organs being registered as representations in the sensory input circuit. But since configurations of qualia are immediately present, they are like objects presented by a faculty of intuition, and so this mistake can be called the *illusion of intuitionism*. It depends not only on many qualia of various kinds all being immediately present at once but also on sensory qualia being configured in phenomenal space as they are in mammals. That combination of properties is called the *unity of consciousness*, and it is what causes the illusion of intuitionism and enables metaphysical arguments to cause a third stage of spiritual evolution.

(2) How Consciousness Helps Cause a Stage of Evolution

The illusion inherent in consciousness is essential to this explanation of the metaphysical stage of evolution because, as ontologists will recognize, consciousness treated as something distinct from psychological states in the mammalian brain is not an efficient cause. That is, consciousness as the immediate presence of phenomenal properties does not make any difference

---

perceive the natural world. How ions are entangled with one another depends on the quantum structure of its mechanical system of the inertial system constituting them, and when the motion and interaction of ions are not constrained by the structure of a geometrical cause, the ions are not entangled in the unique way that causes the holistic power of matter to be expressed as the unity of consciousness, that is, as magnetic field matter whose phenomenal way of existing in itself is a complex phenomenal property that is immediately present to the mammal. Ions in the plasma are entangled, since all particles are entangled as an expression of the non-local power of space. But the ions in the plasma are not entangled in the way that causes magnetic field matter waves to have the kind of complex phenomenal property that mammals have when they perceive the natural world because that depends on the mammalian neural BGS as a geometrical cause and geometrical causes depend on the quantum structure of the mechanical system. Even though it must be like something to be each part of the magnetic field matter helping constitute the electromagnetic system of the plasma, their intrinsic phenomenal properties are not all immediately present to the plasma at once, not even momentarily. What it is like to be the entire plasma lacks the unity of consciousness, and so there can be no illusion inherent in it.

in what happens in the mammalian brain. That is *just* what it is like to be some of the matter helping constitute the geometrical cause that constrains the collective effects of physical causes in the mammalian brain. The phenomenal expression of the holistic power of matter is a previously unrecognized *way of existing*, not a previously unrecognized *way of determining what happens*. Every event is completely determined by geometrical and physical causes. Though the causal impotence of consciousness as something different from psychological states may seem to make it irrelevant in explaining how metaphysics causes a stage of spiritual evolution, it is essential.

Consciousness is not irrelevant because, as ontological scientists will recognize, the illusion inherent in consciousness is what enables metaphysicians to discover that they are conscious. Indeed, historically, that is how most reflective subjects have learned that they are conscious. The immediate presence of phenomenal properties has no effect on the behavior of mammals, not even the behavior of language-using mammals. But when their immediate presence is taken to be the cause of knowledge about them, phenomenal properties are, in effect, separated from the psychological (brain) states in which they inhere, so when metaphysicians refer to them as something different from psychological states, consciousness is incorporated as part of the level of linguistic organization that causes a third stage of spiritual evolution. The belief about its causal role is false, and since it is part of another false belief, there is a mistake that intuitionists can correct without giving up intuitionism. It is a mistake about the nature of perception, and correcting it is how metaphysicians discover that they are conscious. Remarkably, however, even then, the immediate presence of phenomenal properties makes no difference in the geometrical action of the neural BGS because they discover that they are conscious from what they learn about the natural world. And since this discovery about perception does not correct the false belief at the foundation of their metaphysical arguments, they discover that they are conscious in the problematic form of mind-body dualism. That is the intractable puzzle that dooms modern philosophy, and since it is an ontological problem, it is not solved until ontologists reduce physics to spatio-materialism.

Since consciousness is what it is like to be a mammalian brain, it is an ontologically necessary part of metaphysical arguments. But since phenomenal properties are never anything more than objects of reference, they play a strictly passive role, so ontological scientists will call consciousness a *pseudo-efficient cause*. This section describes what ontological scientists will discover about the role of consciousness in causing the metaphysical stage. There is a regularity about how the metaphysical stage unfolds by which they will confirm that Western civilization is the metaphysical stage of spiritual evolution on Earth. That explanation of the West will solve the problem of mind-body dualism ontologically, and its discovery will turn ontological science into naturalistic Reason.

According to the reproductive mechanism, another stage in the evolution of geometrical causes is inevitable whenever a higher level of geometrical organization is both possible and functional. Spiritual organisms are the form of life that evolves on the fourth floor of the edifice of life, and since they have a plan-sharing BGS based on the use of language, the spiritual BGS itself is another embodiment of the reproductive mechanism. Linguistic representations are double-action geometrical causes that go through reproductive cycles as they are exchanged, and as we have seen, linguistic representations evolve by *judgmental selection* because agreement is reached by arguments that lead members to make the same judgment. Evolution during stages of spiritual evolution is mainly cultural, and since ontological scientists will know how the first two stages were caused by levels of linguistic organization, their explanation of consciousness and the illusion inherent in it will enable them to discover how the exchange of metaphysical arguments causes a third minor stage of spiritual evolution.

During the first stage, the linguistic representations are naturalistic sentences (with a simple subject-predicate or subject-predicate-object grammatical structure), and members share plans by following the commands of a leader. They share certain goals, but since they cannot explain their behavior by its goals, what they argue about is means to ends that are merely implicit, such as better ways of making and using tools. Culture evolves by *technical selection*, and the range of functional powers that can evolve is limited.

During the psychological stage, members have the use of psychological sentences (which describe subjects as having psychological states about states of affairs identified by naturalistic and psychological sentences), and this grammatical structure is a higher level of linguistic organization than naturalistic sentences. Mammals with the use of psychological sentences can represent their psychological states as causes of their behavior—or as causes of their beliefs—in the very process of causing them. The goals that reflective subjects use to explain their behavior and the behavior of others are purposes, and since they know which goals others share, arguments about ends and means lead to shared purposes. They argue about which plans to follow, and since linguistic exchanges end when they agree about *what works* in the relevant situation, culture evolves by *pragmatic selection*. The ability to cooperate in pursuing shared goals opens up a range of functional powers that only psychological-level spiritual organisms can evolve.

Though the naturalistic and psychological linguistic levels are evident in the grammatical structure of the sentences introduced, ontological scientists will know that the ability to use them is acquired by genetic changes in the mammalian neural BGS. But metaphysical arguments do not require sentences with a higher level of grammatical structure than psychological sentences. No genetic change in the brain is required for the third level of linguistic organization because it is just the exchange of arguments that refer to consciousness as something different from the states of the mammalian brain in which they inhere. Instead of using a level of organization in the structure of the brain to explain the level of linguistic organization that causes the third minor stage of spiritual evolution, ontological scientists will use the way that intuitionistic metaphysicians refer to consciousness to explain how consciousness helps cause the metaphysical stage.

Mammals are animals that use a faculty of imagination to understand sensory input, and since they are conscious, the unity of consciousness makes it seem that everything that exists is somehow immediately present. That is, mammals are *inside consciousness*, which makes it seem that all knowledge depends on faculties of intuition. Knowing that there is an illusion inherent in consciousness, ontological scientists will understand how metaphysical arguments can be a third level of linguistic organization, as if they were a level of grammatical structure higher than psychological sentences. Metaphysicians hold that Reason knows Reality behind Appearance, and since phenomenal properties are immediately present, the illusion of intuitionism allows them to explain Reason and Appearance as faculties of intuition.

All mammals naturally assume that perceptual phenomenal appearances *are* the natural world, and this does not change when mammals acquire the use of naturalistic sentences. The difference is that the phenomenal appearances of meanings of naturalistic sentences constructed in naturalistic imagination seem to be superimposed on the natural world. They are *naturalistic states of affairs*, which can be seen as causes and effects. And when mammals acquire the use of psychological sentences, meanings of psychological predicates constructed in psychological imagination have phenomenal appearances, and they seem to be the psychological states that they represent in the subjects of which they are predicated, either themselves or other subjects. In short, reflective subjects are naturally *direct realists* about both *perception* and *reflection*.

Knowing what it is like to be a reflective subject, ontological scientists will see metaphysical arguments as a way of using language that separates references to phenomenal appearances from the representations in faculties of imagination that determine their structures. When metaphysicians describe perceptual phenomenal appearances as *Appearance*, they treat (at least part of) consciousness as an object presented by a faculty of perceptual intuition, and since that describes phenomenal appearances as playing a causal role in knowledge that they do not have, it separates them from the states of the mammalian brain that actually play that role and from the bit of matter whose intrinsic property is the phenomenal appearance. Metaphysicians hold that Reason knows Reality behind Appearance, and when they refer to meanings of words or sentences constructed in faculties of imagination as objects of a faculty of rational intuition, they likewise describe reflective phenomenal appearances as having causal roles that they do not have, separating another part of consciousness from the psychological states of the mammalian brain that actually play those causal roles.

Remarkably, consciousness still is not an efficient cause of anything that happens in the brains of metaphysicians. Metaphysical arguments use a false assumption about consciousness to explain their knowledge, but the geometrical structure of the brain continues to constrain the collective effects of physical causes in the same way. The only difference is that metaphysicians, assuming they know about phenomenal appearances because they are immediately present, make arguments that refer to phenomenal appearances in a way that implies that they are different from the representations in their faculties of imagination that determine their structures and actually play the causal roles attributed to them.

Nevertheless, articulating the intuitionist illusion in this way causes a third stage of spiritual evolution because unlike reflective subjects during the psychological stage, metaphysicians do not stop arguing when they agree about how to pursue goals. Their belief that Reason knows Reality behind Appearance motivates them to continue arguing until they agree about how all their shared beliefs about Appearance correspond consistently to the Reality that Reason seems to discover behind it. They expect the nature of Reality to explain what is found in Appearance completely enough for them to know not only what is good but also the nature of goodness. This correspondence theory of truth imposes a higher standard of coherence on the acceptance of conclusions than pragmatic selection. Thus, ontological scientists will know that when metaphysical arguments are tried out as a cultural variation during the psychological stage of spiritual evolution, culture will evolve by *Rational selection* and spiritual organisms will acquire an entire range of new functional powers.

Consciousness is an ontologically necessary part of the cause of the metaphysical stage because mammals are necessarily conscious. But if this were the only role that it played, even metaphysicians would not know that they were conscious because they would still be *inside* consciousness. They would be taken in by the illusion of intuitionism, so everything to which they referred would be a phenomenal property that seems to be known by its immediate presence, including the Reality that they believe exists behind Appearance.

Nevertheless, the exchange of metaphysical arguments leads reflective subjects to discover that they are conscious because the metaphysical stage begins with another false belief that intuitionists can correct without seeing through the illusion of intuitionism. Metaphysicians are initially *direct realists*, but when intuitionists give up direct realism, they become *representative realists*. The study of how objects in space cause perception in animal bodies enabled naturalists to convince intuitionists that objects in the natural world exist independently of one another, and that posed a puzzle. Though such objects in space might have phenomenal intrinsic properties, they could not be the qualia that are immediately present to a mammal perceiving them. Since they exist outside one another (and most of them outside the perceiver's body), they keep to themselves whatever they are in themselves. Thus, when intuitionistic metaphysicians were forced to recognize that the qualia objects in phenomenal space that are immediately present in perception are different from the objects that exist outside their bodies that they seem to represent, they discovered that the natural world is *outside* consciousness, and that is how they discovered that they are conscious.

But to discover consciousness in this way is to discover it in the problematic form of the belief in mind. Modern metaphysicians are still inside consciousness where qualia objects in phenomenal space are all immediately present, and by recognizing that there is a world outside consciousness where their body and other objects exist independently of one another, they learn that the mind has a basically different nature from the body. Metaphysicians can describe the unity of mind because language-using mammals naturally assume that they know about qualia objects in phenomenal space because they are immediately present, and they can describe the unity of the external world because they know that all the objects in it exist independently of one another in three spatial dimensions. Though objects in the natural world might have phenomenal intrinsic properties, it is ontologically impossible for those properties to be immediately present to the perceiver, so metaphysicians know that the body is essentially different from the mind. That poses a problem about how mind and body interact.

This way of discovering that they are conscious does not depend on consciousness causing anything to happen beyond what is caused by mammalian brain states. It all depends on learning about the nature of the external world, and intuitionists learn that by giving up direct realism in favor of representative realism. Consciousness is the immediate presence of phenomenal properties, and though metaphysicians explain their knowledge of those properties as an effect of their immediate presence, that just articulates what mammals believe because they are naturally taken in by the illusion inherent in consciousness. The brain is conscious, but since consciousness itself is not an efficient cause of the discovery that reflective subjects are conscious, ontological scientists will call consciousness a *pseudo-efficient cause.*

Furthermore, the discovery of consciousness in the problematic form of mind-body dualism is, historically, how reflective subjects learned that they are conscious. No such discovery could be made by reflective subjects during the psychological stage. They are taken in by the intuitionist illusion, and as direct realists about perception, they assume that they know about the natural world because it is immediately present. But since they do not *describe* their knowledge of phenomenal properties as effects of their immediate presence, they cannot even conceive of representative realism. They are trapped inside consciousness.[23]

Knowing that the exchange of metaphysical arguments is what leads to the discovery of consciousness, ontological scientists will use the history of philosophy to confirm that Western civilization is the metaphysical stage on Earth. It began in ancient Greece and eventually became modern metaphysics, and since modern culture continued to evolve by Rational

---

[23] Though Zen meditation makes one focus on the essential nature of conscious, it is known by the way that it differs from the complex phenomenal properties of ordinary life that cause suffering, not by how it differs from the natural world.

selection, its confidence in Reason brought physical science into existence. The remainder of this chapter sketches how these metaphysical developments led to the 18th-Century Enlightenment, explains how its aftermath eroded the third level of linguistic organization, and predicts that the completion of the First Enlightenment will lead to the discovery that Western civilization is the metaphysical stage. That is the Second Enlightenment. It turns science into naturalistic Reason serving its theoretical function in a perfect way, and after showing how naturalistic Reason also serves its practical function in a perfect way, the final chapter speculates about where it leads in the end.

(3) Ancient Metaphysics

Plato was the intuitionist who discovered Reason. He made the first attempt to show that beings like us have a cognitive capacity that enables us to know Reality behind Appearance. By *Appearance* he meant perceptual phenomenal appearances, and by analogy to how they seem to give us knowledge of the natural world, he assumed that certain reflective phenomenal appearances are the Reality that Reason knows behind them. In both cases, Plato was taken in by the illusion of intuitionism. He was a *direct realist about perception* because, as a mammal, he believed that perceptual phenomenal appearances *are* the natural world. And he can fairly be called a *direct realist about Reason* as well because he believed that certain reflective appearances *are* the Reality that Reason knows behind Appearance. Assuming that the immediate presence of perceptual phenomenal appearances is what makes them known to us, he described perception as a faculty of intuition by which we know about visible objects that exist in the *realm of Becoming*. And assuming that the immediate presence of certain reflective phenomenal appearances is what makes them known to us, he described Reason as a faculty of rational intuition by which we know about abstract objects that exist in the *realm of Being*. He called the latter *Forms*, since they were the reflective phenomenal appearances of the meanings of general terms, such as *equality*, *circle*, *tree*, and *virtue*. As a metaphysician, Plato held that Forms are the Reality behind Appearance. He believed that Forms were what really exist because they were unchanging, and he insisted that visible objects were a mere Appearance of Reality because they were just changing imitations of Forms.

Though the realm of Becoming was just a kind of phenomenal appearance, it was caused by a world that really exists. But Being was just a phantom Reality because without the illusion of intuitionism, there would be no such realm, and so the assumption that Being and Becoming both exist caused an intractable puzzle. For the Reality discovered by Platonic metaphysics to explain everything, Forms would have to explain not only all the visible objects *in* Appearance, but also Appearance itself, and though ancient metaphysicians were slow to recognize it, pre-Socratic philosophers had set an exacting standard for an explanation of everything. The coherence of an ontological explanation requires a first cause, and if Reality is the first cause, the substances constituting Reality must constitute everything in the world, including Appearance itself. But unchanging Forms have an opposite nature from changing visible objects, so no substance that constitutes Forms could also constitute visible objects. The essential nature of a substance is how it exists in itself, and change cannot be constituted by a substance whose way of existing is not changing, any more than a circle can be a square. The dualism of Being and Becoming was the ontological problem that doomed ancient metaphysics.

But the problem was caused by the belief in a phantom Reality. Plato's intuitionistic metaphysical argument conjured up a phantom Reality because the objects known by rational intuition were supposed to exist in a realm of Being outside space and time and nothing exists outside the natural world. To be sure, neither do objects of intuition exist. But they are not a phantom Reality because phenomenal properties really exist and it is just their nature that is misunderstood. Nor is the realm of Becoming a phantom Reality. It is just a misunderstanding about the nature of a world of objects that really exist in space. The realm of Being is a phantom Reality because it cannot be constituted by any substance whose endurance through time enables it to constitute change, like those that constitute the natural world.

Believing that Reason can know Reality behind Appearance, ancient metaphysicians tried to defend Plato by showing how beliefs are true by correspondence to Forms, and Western culture continued to evolve by Rational selection. The challenge was to find a coherent ontological explanation of the dualism caused by Plato's phantom Reality, and since that was impossible, they wound up trying to hide Plato's dualism. There are only two ways to hide the ontological dualism caused by intuitionistic metaphysics; both were offered as explanations of the relation between Forms and visible objects; and both failed the test of ontological coherence.

Aristotle defended a teleological naturalism (called *hylomorphism*) in which every object in space was a combination of matter and form. Matter was supposed to enable objects in space act on one another as efficient causes, while essential forms were supposed to be responsible for their *natural change*, that is, their change in the direction of an end called the *final cause*. But locating the Platonic dualism in particular objects did not make the combination of their incompatible natures any more ontologically coherent. It was still not possible for two opposite properties to be how substances exist in themselves.

Much later, Plotinus defended an ontology in which all the Forms and all the visible objects were described as objects that are immediately present to a single subject called *The One*. This kind of ontology, called (*absolute*) *idealism*, explains everything by a substance with an intuitionistic nature. It takes Reality and Appearance to be the same thing, and since it gives up the claim that Reason knows Reality behind Appearance, it replaces Truth as correspondence to Reality with the theory that Truth is the coherence of its explanation of everything. Idealists may claim to have a first cause, but their metaphysics fails the pre-Socratic test of ontological coherence because it is not possible to understand in naturalistic imagination how intuitionistic substances constitute objects that actually exist independently of one another in space like those in the natural world.

(4) Modern Metaphysics

If Plato was the intuitionist who discovered Reason, Descartes was the metaphysician who used it to discover consciousness. Metaphysical-level culture evolves by Rational selection, and when ontological scientists explain the history of philosophy, they will discover how confidence in Reason and the exchange of metaphysical arguments in Western culture led to the discovery of consciousness. In the modern era, nearly two millennia after Plato, there were so-called *natural philosophers* defending the ontological naturalism from which empirical lawism would blossom, and they recognized that the only way to perceive objects that exist independently of one another in space is by their effects on animal bodies. Even intuitionists had to reject direct realism about perception in favor of representative realism, and this spurred Descartes to introduce a new kind of metaphysical argument that revealed the existence of consciousness, albeit in the problematic form of mind-body dualism.

Descartes intended his metaphysical argument to give natural philosophy a foundation that was known with certainty. As an intuitionist, he believed that perceivers use phenomenal appearances to know about the natural world, and recognizing that perceptual phenomenal appearances are parts of the perceiver, he gave up Plato's direct realism in favor of *representative realism*. But representations are different from what they represent, and as a metaphysician, he had to prove the existence of the world external to the perceiver that they represent. That is what led to the discovery of consciousness. Since Descartes assumed that he knew about phenomenal appearances because they were immediately present, he still mistakenly assumed that perception depended on a faculty of intuition. He called the Reality that Reason knows behind the perceptual phenomenal appearance the *external world,* and assuming that Reason knows Reality behind Appearance, Descartes tried to prove its existence. His arguments depended on showing that the God who created the natural world could not be a deceiver, and the fallacies in those arguments are now taught as lessons in critical thinking. But the Achilles Heel of intuitionistic metaphysics was the ontological incoherence of its explanation of everything. The external world was not a phantom Reality because the natural world does really exist. The incompleteness of Descartes' description of its nature was the mistake. But since he used perceptual phenomenal appearances to prove its existence, he had also to explain the existence of phenomenal appearances of all kinds, and his way of showing how Reason knows the Reality behind them turned up the problematic ontological dualism that doomed modern metaphysics.

Intuitionists assume that knowledge of phenomenal appearances is caused by their immediate presence, and calling the knower *mind,* Descartes described phenomenal appearances as *ideas* in the mind. There were some reflective phenomenal appearances that he could treat as objects of a faculty of rational intuition. They were not unchanging objects like Platonic Forms but, rather, arguments that seemed self-evident because they were arguments based on the structure of a faculty of imagination built into the mammalian neural BGS. Believing that their immediate presence made phenomenal appearances indubitable, Descartes called them *clear and distinct ideas,* and as usual, the paradigm of knowledge by rational intuition was mathematics. It explained, for example, how we know that a triangle has three sides and that the sum of their interior angles equals two right angles. But the conclusion he drew about the nature of the external world was incomplete. Assuming that mathematics was known by a faculty of rational intuition built into the mind, he believed that the external world had to have the essential nature described by mathematics, and thus, he concluded that the external world was a substance spread out in three dimensions like space, called *extension.* To explain how it could change as time passes, he had to assume that its parts were able to move relative to one another, and thus, extension was necessarily divisible into parts. Since this was to overlook the difference between matter and space, it was incomplete. But it did imply that objects in space exist independently of one another, and that is, indeed, part of the essential nature of the natural world.

Since mathematical ideas enabled him to demonstrate the existence and nature of the external world, Descartes had to explain the existence of ideas. As a metaphysician, he assumed that Reason knows Reality behind Appearance, and he found another clear and distinct idea in the mind that enabled him to show the existence and nature of the Reality behind phenomenal appearances of all kinds. It was an argument, like mathematical arguments, that seemed to be self-evident. Descartes' argument was *"I think, therefore I am."* Intuitionists could not doubt this argument because it was just a description of the intuitionist illusion. *Thinking* was the immediate presence of phenomenal properties (ideas generally), and since that required a subject to whom they were present, there had to be a thinker. This famous argument, called the *Cogito,* was so compelling that Descartes introduced it by arguing that it is what anyone who tries to doubt everything must believe. It was his standard for declaring that all *clear and distinct ideas,* including mathematics, were indubitable. But as ontological scientists will discover, the *Cogito* is dubitable. The use of psychological sentences gives mammals the faculty of imagination by which they can reflect on their psychological states, and that cognitive power explains the indubitability of the inference from *I think* to *I am*. Though reflection on thinking has a phenomenal appearance, intuitionists are mistaken in believing that the *Cogito* is indubitable because what is immediately present is self-evident.

Since Descartes relied on a faculty of rational intuition to show the existence and nature of both mind and the external world, modern intuitionistic metaphysics encountered an ontological problem about dualism that was as unsolvable as the one that Plato encountered in the ancient era. Descartes' argument about the nature of the external world implied that it was divisible into parts, and since he used the *Cogito* to describe the nature of the mind, he knew that mind had a unity that precluded division into parts. Ideas in the mind are phenomenal properties in which many qualia of various kinds are all immediately present to a subject at once, and since they could not be in the mind without being immediately present, qualia could not exist separately from one another like parts of extension. To be sure, the external world also had a unity. Even when its parts were separated from one another, the places where they existed were all related in three spatial dimensions. But the body could not be identical to the mind because it was part of extension, and if parts of the brain had phenomenal ways of existing in themselves, that was something they had to keep to themselves. They could not be immediately present to a single subject all at once. Since the mind had a unity that was just opposite to the unity of extension and the body, the existence of both mind and body was an ontologically problematic dualism.

Substances constitute Reality, and having shown that mind and body were both parts of Reality, Descartes had to show how both kinds of substances were parts of the same world. Any ontologically coherent explanation of everything in the world would have to explain how mind and body interact, and that was not possible. The unity of mind and the unity of the body entailed basically different kinds of efficient causes. The immediate presence of phenomenal properties is what caused the mind to hold beliefs and make choices, while the behavior of the body and everything else in the natural world was caused by parts of extension moving relative to one another. Mind-body dualism was the ontological problem that doomed intuitionistic metaphysics in the modern era. But it will be solved when ontological scientists explain Western civilization as the metaphysical stage because they will discover that the mind was just the phantom Reality conjured up by Descartes's metaphysical argument in much the same way that Being was the phantom Reality conjured up by Plato's metaphysical argument in the ancient era.

It is noteworthy that mind-body dualism was not the only ontological problem confronting Cartesians. Mind and body were both substances constituting the natural world, but since Plato's belief in a realm of Being had evolved into the belief in a God who created the natural world (more about that in the next chapter), a world containing both mind and body was just one part of the ancient problem of ontological dualism. This compounded ontological dualism will also be solved because when ontological scientists explain Western civilization as the metaphysical stage of evolution, they will discover that mind and God are both phantom substances caused by the mistaken belief about knowledge on which intuitionistic metaphysics was based. Though both eras of intuitionistic metaphysical arguments depended on consciousness, Plato did not discover consciousness because when he discovered Reason, the Reality that he discovered was inside consciousness along with both Reason and Appearance. Despite having opposite natures, Being and Becoming were both parts of consciousness. The new kind of metaphysical argument in the modern era was also formulated inside consciousness. But Reason knew about a world outside consciousness, and since the divisibility of the natural world was opposite to the unity of consciousness, it was the discovery of consciousness.

Notice once again that consciousness itself was not an efficient cause of Descartes's knowledge that he was conscious. To be sure, that is what he assumed by using the *Cogito* to show that Reason knew the Reality behind consciousness because it described the immediate presence of phenomenal appearances as causing knowledge of them. That is how Descartes discovered consciousness in the problematic form of the belief that mind is a substance with a nature opposite to body. It was a phantom Reality. The illusion inherent in consciousness enabled him as a metaphysician to refer to consciousness as something different from the brain, and since that caused his belief in mind, consciousness itself played a strictly passive role. It was just a *pseudo-efficient cause*.

This explanation of the origin of the belief in mind implies that minds are identical to the brains of reflective subjects who explain their knowledge of consciousness as an effect of the immediate presence of their phenomenal properties. That is, minds are identical to brains of intuitionists. There have always been naturalists, such as Hobbes, who assumed that the mind is just the brain. But materialists cannot solve the mind-body problem because its solution depends spatio-materialism. To explain mind, ontologists must reduce physics to ontology, unify science, and use spatio-materialism to explain how consciousness is part of the natural world. Moreover, there have always been philosophical skeptics about the belief in mind as well. But they were skeptical about the phantom Realities conjured up by metaphysics, and since they never doubted intuitionism, they became epistemologists, like the British empiricists, whose goal was to explain how science is possible without having to show that Reason knows Reality behind Appearance. But the motor of philosophy was metaphysics, and as culture continued to evolve by Rational selection, it was taken for granted that truth depends on correspondence to Reality.

In the modern era, metaphysicians tried to solve Descartes' problem of mind-body dualism by showing how mind and body could interact. But as in the ancient era, the most that they could do was hide Cartesian dualism in another kind of ontological theory. Spinoza, like Aristotle, tried to show that *Thought* and *Extension* were two attributes of a single substance that constituted everything that exists. That did not pass the pre-Socratic test of ontological coherence because the essential nature of a substance is how it exists in itself and no substance can have both the unity of mind and the divisibility of extension. And Leibniz, like Plotinus, tried to reduce Reality to Appearance. Leibniz postulated many finite mental substances called *monads*, and since their perceptual phenomenal properties all fit together consistently as if they had bodies that were related spatially as parts of a natural world, it was the reduction of the natural world to perceptual phenomenal appearances. This abandoned the basic metaphysical claim about Truth being correspondence to Reality, and it did not pass the Pre-Socratic test of ontological coherence because monads could not constitute objects in space. However, Spinoza and Leibniz both tried to solve the ancient as well as the modern ontological problem. Spinoza explained God as a single substance in which every idea in the attribute of Thought corresponded to a body in the attribute of Extension, including some ideas that were minds corresponding to bodies of animals. And since Leibniz explained God as an infinite monad that created the finite monads, he explained God's relation to them as the cause of their being the best of all possible worlds, or perfection.

There was no adequate solution to mind-body dualism, and believing that modern metaphysics was bound to fail, Kant denied that it was possible for Reason to know the *thing in itself*, the Reality behind Appearance. But he defended a constructivist theory of Reason that showed how physics could discover universal laws of nature that hold necessarily, and as we have seen, his explanation of mathematical truth paved the way for epistemologists to explain the possibility of science without knowing Reality behind Appearance. The use of mathematics to formulate laws of nature was the secret sauce of physics, and by reducing mathematics and laws of physics to interactions of space and matter, ontological scientists will be able to trace the origin of their own complete explanation of nature to the physical science spawned by the Enlightenment and discover that metaphysics was the origin of the secret sauce of physical science. But having shown that the illusion of intuitionism caused the mind-body problem and other philosophical problems, ontologists will see Kant's defense of the Enlightenment as the death knoll for intuitionistic metaphysics, and in a culture where mind-body dualism is how reflective subjects learn they are conscious, ontologists will expect the failure to distinguish between consciousness and reflection to cause profound confusion. The plight of science as well as philosophy will be desperate, and since that is the historical context in which ontologists will be explaining the origin of ontological science, they will see what happened after the 18th-Century Enlightenment as confirming their explanation of Western civilization as the metaphysical stage of evolution.

(5) The Aftermath of the Enlightenment

The aftermath of the Enlightenment saw the petering out of confidence in Reason. While belief in God persisted as the confession of faith in a religion, theoretical Reason continued to evolve. It evolved along two different pathways, one as philosophy, and the other as science. But neither will be the kind of cultural progress expected during the Enlightenment, at least, not until ontologists reduce physics to ontology and unify science. This chapter shows how their discovery of the identity of mind and brain will unify science and philosophy and vindicate belief in theoretical Reason. The next chapter predicts that a parallel perfection of practical Reason will vindicate the belief in God.

Ironically, in the late modern era, Kant's argument about the impossibility of Reason ever knowing Reality behind Appearance provided an opening that enabled intuitionists to renew their defense of metaphysics. The *Critique of Pure Reason* entailed a dualism of phenomenal and noumenal worlds. One was the world that Reason helped construct, while the other was the world of things in themselves that Reason could not know, and modest modifications of Kant's constructivist theory of Reason enabled metaphysicians to hide Kant's problematic dualism. They used the same two strategies as those used in the ancient and modern eras: a double-aspect theory, and a form of idealism. Both attempts postulated something called the *Absolute*, and attributing to it constructivist powers like those that Kant had used to explain Reason, they treated the Absolute as the Reality that Reason knows behind Appearance. Schelling played the role that Aristotle had played in the ancient era and that Spinoza had played in modern era by arguing that mind and body were, respectively, the *subjectification* and the *objectification* of the Absolute. Hegel played the role that Plotinus had played in the ancient era and that Leibniz had played in the modern era by arguing that the Absolute was Reason with a *dialectical* nature that enabled it to discover itself as Reality behind Appearance. Neither passed the pre-Socratic test of ontological coherence. The former was incoherent because it implied that the Absolute caused both mind and body to exist, and it is not possible for a substance to constitute entities with two such opposite natures. The latter was incoherent because if Reason were the Absolute, it would be an intuitionistic substance, and it is not possible to understand in naturalistic imagination how it constitutes a world in which objects exist independently of one another in space.

Hegel's metaphysical argument set the tone for academic philosophy in the 19th-Century. Intuitionists defended various other versions of *absolute idealism*, and since they reduced Reality to Appearance, their theories could not explain the True as correspondence to Reality behind Appearance. So, they replaced the correspondence theory of truth with the *coherence theory of truth* and defended the claim that Reason knows *the True* on the grounds that it had a coherent explanation of everything as something that is immediately present.

Philosophers were eventually forced to abandon metaphysics because idealism was not credible in light of scientific advances on the other side of the abyss. Scientists are naturalists, who assume that the natural world is what really exists, so they use the empirical method to learn about it. Since the astonishing success of physics in discovering basic laws of nature led naturalists to believe that everything is constituted by matter, those who expected to know the True had to choose between materialism and idealism. Either the world was constituted by material substances, or else it was a single intuitionistic substance. There was, however, one issue on which science and philosophy agreed. Both rejected supernaturalism. Neither needed a transcendent God. Indeed, this consensus forced those who believed in God to admit that it was a matter of faith, not the conclusion of argument. Since the naturalism shared by science and philosophy was simply the abandonment of the modern remnant of ancient ontological dualism, it was a basically negative agreement. But there was no way to reconcile their other views about what exists, and as the abyss between them grew, science and philosophy both lost confidence in Reason.

Science prospered from the spectacular success of physics in discovering basic laws of nature, and since its mathematically formulated laws could be interpreted as descriptions of powers of material bodies located in absolute space, it was seen as justifying materialism. Though the rejection of a basic assumption of intuitionistic metaphysics was the beginning of physics, it did not reject intuitionism completely.

Naturalists are ontologists because they assume that the natural world is what exists, and as naturalists, scientists assumed that their conclusions about the natural world were confirmed by measurements, that is, by sensory input to their animal bodies. That was to reject the assumption that perception depends on a faculty of intuition, and since that was a basic assumption of intuitionistic metaphysics, the difference between science and philosophy might seem to be the difference between naturalism and intuitionism. But naturalism is an ontological assumption, and ontologists will not be taken in by that simplistic view of the abyss between science and philosophy.

Ontological scientists will know that the belief in a priori mathematics was responsible for the success of physics. Metaphysicians had always used mathematics as the prime example of knowledge that did not depend on perception, and Kant used a constructivist theory about the truth of mathematics to explain how mathematics could be known *a priori* even though he denied that Reason could know Reality behind Appearance. But ontological scientists will know that belief in *a priori* mathematics kept physics from discovering the role that space plays as a substance in helping constitute regularities in the natural world, and though the success of physics inspired the use of the empirical method in studying all kinds of regularities found in the natural world, they will also know that this taint of intuitionism in physics crippled the specialized sciences. It deprived physical science of the second kind of efficient cause that was needed to explain the regularities they found in the study of nature. Expecting physical science to be crippled by physicalism, ontologists will see the four famous scientists commonly used to illustrate advances in scientific understanding as the *four horsemen of disillusionment with Reason*.

In biology, Darwin's explanation of the evolution of species by natural selection replaced a purposiveness in nature that seemed to have a divine origin with an accidentalism about its course that disabused us of the belief in an objective difference between good and evil. To this day, scientists still warn us that science cannot tell us what we ought to choose.

In economics, Marx defended the discovery of a motor of history that explained the rise of capitalism as just the replacement of feudalism by another form of class oppression. In his view, the Enlightenment belief in the moral autonomy of Rational beings was just an ideology whose function was to justify capitalist exploitation of workers, much as the Christian view that everyone was created in God's image was just an ideology whose function was to justify lords exploiting serfs. But Marx expected a vanguard of scientists who understood the motor of history to see through liberal ideology and spark a revolution in which the free market would be replaced by socialism, that is, by government ownership and management of the means of production.

In psychology, Freud held that while the ego follows the reality principle and plans behavior to maximize satisfaction of desires, choices between good and evil depend on the ego being forced by fear of punishment to submit to the dictates of a superego internalized from parents. Freud's explanation of morality meant that the human condition was having to suffer what he called the *discontents* of civilization.

And in physics, Einstein held that Newton was mistaken in believing that space and time are absolute. That is how they are understood in naturalistic imagination, and his special theory of relativity replaced them with spacetime, an entity in which all moments in time exist in the same way, like all the points on a line. Spacetime cannot explain our belief that existence is what distinguishes the present moment from all other moments in time. Physicists nevertheless insist that spacetime is what really exists because surprising measurements of many different kinds have confirmed the general theory of relativity, that is, Einstein's explanation of the gravitation by the curvature of spacetime, in such a compelling way.

These scientific theories are commonly used to teach lessons about the hubris of believing that beings like us are capable of the perfect kind of knowledge expected of science after the Enlightenment, promised by metaphysics for millennia, and first sought by the pre-Socratics in ancient Greece. Though confidence in Reason had inspired hope of progress in culture, the pathway of science was marked by the milestones in disillusionment with Reason laid down by these four horsemen.

On the other side of the abyss, however, intuitionism took over completely. Though progress in science was limited, metaphysicians defending absolute idealism were embarrassed by their inability to keep up with science and the unprecedented technological control over nature it spawned. And the intractability of the Cartesian mind-body problem caused even philosophers to deny the possibility of the perfect knowledge that they had once claimed for Reason,

Confidence in Reason came from metaphysics, and as the coherence criterion of truth lost credibility, the first step away from metaphysics was *romanticism*, the attempt to replace Reason with emotion. It was not hard to despise Reason after the Enlightenment because the most prominent evidence of the progress it promised was the industrial revolution. In rural regions, the growth of capitalism was an assault on the remains of the feudal system, while in cities, it replaced traditional aristocratic aesthetic standards with utility. Romantics were offended. But they did not give up the metaphysical aspiration to be in touch with Reality behind Appearance. Instead, they believed that intense feelings, ranging from awe at the sublime to horror at ghosts, were a way of transcending the ordinary world of Appearance and having an emotional relation to something more real behind it. Other romantics found transcendence in acts of will, ranging from individual acts of heroism to the worship of a culture expressed as group action. Hence, romantics were inclined to defend passionate collective action in the name of Reason, as in the Jacobin reign of terror during the French revolution shows.

Without knowledge of the True, however, there is no knowledge of the Good. It was hard for intuitionists to avoid the conclusion that there is no objective difference between good and evil to guide our choices, and ontological scientists will see it as causing the anxiety that fascinated *existentialists*. Kierkegaard dealt with this challenge in the 19th century by replacing Reason with a *leap of faith*, made with fear and trembling at its possible costs. It gave one an absolute relation to the absolute in which it became clear what is good. Nietzsche described the discovery that there is no such thing as the Good was the death of God. He hoped to overcome the loss of a standard of goodness determined from outside the natural world by a will to power inside the natural world that would turn man into superman. In the 20th century, the belief that nothing could be known but what is immediately present became an academic school called *phenomenology*. It was the foundation on which Heidegger insisted in *Being and Time* that we must simply accept our thrownness into the world as a way of existing that cares about what happens even though the world itself does not. Phenomenology was also the foundation for Sartre's *Being and Nothingness.* The immediately present is a world of beings with ways of existing in themselves, but since we have a free will, we have no way of being in ourselves. We are like nothingness, except that we bring what goodness there is in the world into existence by choosing it.

Ontological scientists will also see intuitionistic philosophers as taking a more radical step away from confidence in Reason by rejecting even the limited kind of truth that science claimed to have. Though the Frankfurt School was established to defend Marxism, the worker uprising that was supposed to overthrow capitalism had not occurred, and Marx's belief that science could know the motor of history was abandoned in the 1930s when the School was taken over by Horkheimer and Adorno. Though they continued to believe that capitalism or some unknown social structure imposed an ideology on culture, they were intuitionists with only culture to analyze, so they defended a method, called *critical theory,* that was supposed to identify ideological distortions (often with the help of Freudian psychology). It would emancipate people from ideological domination. And it would arouse the same kind of visceral indignation against unequal treatment as Marxism, but without a motor of history or any scientific explanation of ideology to justify their denunciation of the inequality.

As ontological scientists will find, the final step away from metaphysics was taken by intuitionistic philosophers who denied that there is any such thing as the True, and wallowed giddily in the discovery that there is no objective knowledge of any kind. Recognizing that we are cultural beings, Foucault argued that people use language to exert power over other people. And explaining discourse as an interpretation of something said in a context, Derrida used a method called *deconstruction* to show that instead of discovering the True, the use of language is never anything more than someone offering an interpretation of another interpretation. Like critical theorists, they and other *postmodern philosophers* were intuitionists analyzing culture from inside, that is, from the point of view of members of spiritual organisms, and since talk about criteria of truth was just another way of interpreting what is said, their skepticism about science became the defense of relativism about facts and values. Still, they claimed to know how discourse enables some people to exert power over others, and so, like Marxists, they settled self-righteously for arousing visceral anger in listeners.

In sum, across the abyss from science, the decline of the Enlightenment to endarkenment was led by Continental philosophers. Though postmodernism eventually immigrated to America, it replaced a native way of denying that truth is correspondence to Reality whose modesty did not make philosophers giddy. Ontological scientists will find philosophers, like William James and John Dewey, who dismissed metaphysics as a foolish quest for certainty. Since they were intuitionists who rejected the correspondence theory of truth, they defended an idealist ontology. But their idealism was not absolute because they replaced the coherence theory of truth with pragmatism, and this was a palatable substitute for objective knowledge. American pragmatists insisted that all we can know—and that all we really need to know—is what works in local situations. This was a return to the psychological stage of spiritual evolution and cultural evolution by pragmatic selection, in which an optimistic confidence that we share basic goals took the place of religion.

The Continental gospel of relativism was, however, a latecomer to America. In the 20th century, analytic philosophers in the Anglo-American traditional took sides with science across the abyss from the rising tide of endarkenment. The British empiricists had been skeptics about metaphysics, and following their lead, logical positivists tried to show that metaphysics was meaningless. Though they were intuitionists, they stopped basing arguments on the assumption that perception depends on objects in a faculty of intuition (or *sense data*, as they were called), and since they were direct realists about perception, their intersubjective analysis of language seemed to be objective. Though the analysis of language began with the attempt to reduce mathematics to logic, *logicism* failed, and continuing to believe that mathematics is somehow known independently of perception, they remained faithful to empirical lawism. Since they thought of themselves as naturalists, analytic philosophers were allies of science helping solve scientific problems. Since they were intuitionists using faculties of psychological imagination intersubjectively, they saw themselves as under-laborers of science. But they had the function of a Pretorian Guard of the intuitionistic treasure taken from philosophy that had made physics so successful. However, that did not make analytic philosophers relativists about truth. They used the analysis of language to defend the so-called *deflationary theory of truth*, which holds that sentences are true when they correspond to the natural world in the ordinary intersubjective way.

As it became clear late in the aftermath of the Enlightenment that basic values were in dispute, the defense of pragmatism became less optimistic and more realistic. Among Anglo-American philosophers, the later Wittgenstein agreed with postmodernists about our inability to escape from language games. But instead of using the analysis of ordinary language for political purposes, he merely used it to show that intractable philosophical puzzles could be dismissed as linguistic illusions. And Richard Rorty was an intuitionist whose way of giving up the belief that Reason can know Reality behind Appearance was to settle for a way of knowing about Appearance that historically had proved to be reliable. But pragmatism was still a return to the psychological stage, and ontological scientists will see its defense as the attempt to paint a pretty face on relativism.

(6) The Discovery of the Identity of Mind and Brain

In the aftermath of the Enlightenment, as logical positivists defended behaviorism by dismissing talk about mind as meaningless metaphysics, the mind-body problem lurked in the background. But not long after the middle of the 20th century, the problem of mind was taken up by analytic philosophers. As naturalists, they predicted that science would discover that the mind is identical to the brain. But as physicalists, they discovered that defending this prediction was more difficult than they had expected. After exploring issues for two or three decades, they called the challenge they faced the *hard problem of mind*, by contrast to the *easy* problem of explaining how physical causes enable the brain to serve its functions as a bodily organ. Around this time, neural scientists began to recognize that their explanation of the brain would have to include an explanation of mind. Though laborers in science are wary of eager volunteer under-laborers, neural scientists gained respect for philosophers as they found themselves entertaining similarly preposterous theories about what exists to explain mind. To their surprise, the hard problem of mind was not a philosophical puzzle that science could easily solve. Mind was a problem for the under-laborers in science because they had discovered that physical science was unable to explain it. At least, that is how ontological scientists will see it.

Physicalists assume that physical causes are complete and can explain everything that happens in the natural world, including what the brain does. The first step in showing that the mind is identical to the brain is to explain *ideas*, since they are supposed to cause beliefs and choices. But ideas are phenomenal properties, and laws of physics do not mention them because they describe only the motion and interaction of particles. Without a foundation for explaining how phenomenal properties are part of the natural world, the simplest and least intrusive way for physicalists to accommodate them is to assume that brain states cause complex phenomenal properties and stipulate that they have no effect on what happens in the brain. This theory about how consciousness is part of the natural world is called *epiphenomenalism*, and in a science based on empirical lawism, it takes the form of a *psychophysical law*. But this poses another problem that physical scientists cannot solve. They cannot explain *how they know* that they are conscious. Physical properties do not pose this problem because they have effects in the world. But if phenomenal properties have no effect at all on what happens, physical scientists cannot explain how they know about them. To claim that their knowledge of phenomenal properties is caused by their immediate presence would be to give up both epiphenomenalism and the completeness of physical causes.

To be sure, if phenomenal properties are identical to brain states, physical scientists could explain verbal behavior describing phenomenal properties because they would have physical causes in the brain. But since phenomenal properties are epiphenomenal, they do not affect what happens, and the completeness of physical causes implies that the brain would generate the same verbal behavior even if there were no phenomenal properties. Thus, physical scientists could not explain how their knowledge of phenomenal properties is different from their knowledge of brain states, and since they could not explain how they know that they are conscious, their claim that the mind is identical to the brain would be empirically meaningless.

Ontological scientists will see this obstacle as showing the impossibility of a physicalist solution to the problem of mind. It starts with the physical explanation of the brain and, then, tries to explain the mind, and that approach is hopeless because they must explain how they know they have phenomenal properties. But with a monistic ontology, the only way that physicalists can do that is to assume that knowledge of phenomenal properties is an effect of their immediate presence, and since that is how Descartes explained knowledge of ideas in the *Cogito*, they are saddled with the belief in the mind and the problem of explaining how it interacts with the body.

What enables ontological scientists to solve the mind-body problem is their discovery that the natural world is constituted by space and matter. They can explain how consciousness is part of the natural world, and that overcomes the obstacle that stymies physical science. But in using it to solve the mind-body problem, they approach the problem from the opposite direction. Instead of starting with a physical explanation of the brain and trying to explain the belief in mind, ontologists start with the belief in mind and explain the physical science of the body. That is, their ontological explanation of how consciousness is part of the natural world explains how they know that they are conscious, and that reveals an illusion inherent in consciousness that enables them to explain the origin of the belief in mind as an evolutionary development that shows how the mind is identical to the brain.

To recapitulate, their basically historical approach is represented in Figure 10 by the two arrows pointing toward the circle labeled *Philosophy*. The discovery that space is a substance that interacts with matter enables ontological scientists to explain how phenomenal properties are part of the natural world, and since they know that the mammalian forebrain has a structure that serves as a faculty of imagination, they infer that matter has a proto-phenomenal intrinsic property as the best explanation of why complex phenomenal properties are immediately present to mammals. As we have seen, ontologists describe perception as many qualia of various kinds immediately present at once, and since they know that the natural world is constituted by bits of matter that mostly coincide with different parts of space, they recognize that complex phenomenal properties have a unity that is opposite to the unity of the natural world. They know that whatever distinct bits of matter may be in themselves, it is something they keep to themselves. But the magnetic field matter helping constitute the geometrical action of the mammalian brain is a continuously connected bit of matter, and ontological scientists know that every small volume of it has a spatiotemporal structure that can explain the structure of complex phenomenal properties. Knowing that matter can have a proto-phenomenal intrinsic property, they infer, therefore, that consciousness is what it is like to be magnetic field matter when its coincidence with space has the kind of spatiotemporal structure that the mammalian brain imposes on it. This ontological explanation of consciousness is how they know that they are conscious, and it is not problematic because it is just an inference to the best explanation of what it is like to perceive the natural world. It does not depend on the immediate presence of phenomenal properties having any effect on what happens in the neural BGS that is not the effect of the neural BGS as a geometrical cause. But when ontological scientists discover that they are conscious, they realize that they have always been conscious, and since consciousness is just a phenomenal intrinsic property of magnetic field matter helping constitute the geometrical action of the mammalian brain, they recognize that every mammal that has ever lived has been conscious.

This explanation of consciousness is not a solution to the mind-body problem. Though it implies that consciousness is an essential property of the mammalian brain, it does not explain the belief in the existence of mind. But ontologists can explain that belief because their explanation of consciousness reveals an illusion inherent in consciousness. The unity of consciousness makes it appear to mammals that perceptual phenomenal properties *are* the natural world and that reflective phenomenal properties *are* the psychological states on which they reflect. It seems to mammals, therefore, that the immediate presence of phenomenal properties is how they know about objects in space and their feelings about them. This is the *illusion of intuitionism*, and since language-using mammals are inside consciousness in this way, ontological scientists can explain metaphysical arguments as a way of referring to phenomenal properties as something different from the psychological states of which they are parts. That is, when intuitionistic metaphysicians argue that Reason knows Reality behind Appearance, they refer to phenomenal properties in a way that assumes that their immediate presence is what causes knowledge of them, like objects presented in a faculty of intuition. Thus, references to phenomenal properties as Appearance and as Reason incorporate *parts of consciousness divorced from the brain* into a third level of linguistic organization. That is how intuitionistic metaphysics conjures up phantom Realities, and so using the reproductive mechanism to explain Western civilization as the metaphysical stage, ontological scientists discover the origin of the belief in the existence of mind and why it is so widespread—even though it is false. As we have seen, metaphysics begins in the ancient era, and when metaphysicians are disabused of direct realism about perception, metaphysical arguments saddle them with the mind-body problem. The *Cogito* is a description of the intuitionist illusion, and since the illusion never goes away, the belief in *mind* spread widely in the modern era.

This evolutionary explanation of Western civilization is how ontological scientists start with their ontological explanation of mind and, then, explain body. That approach follows their historical order in evolution. Since all mammals are inside consciousness, they implicitly assume that they are minds. But they do not know that they are minds until language-using mammals take up metaphysics and discover a world outside the mind that exists in a basically different way from what is immediately present in the mind. That development during the metaphysical stage of evolution is the discovery that reflective subjects are conscious. Knowledge of consciousness is not the immediate presence of phenomenal properties, as intuitionists assume. It is, rather, caused by recognizing the difference between the familiar unity of consciousness and the unity of a world outside consciousness that is made of substances that exist outside one another in the three dimensions of space. Since reflective subjects are conscious, they have always known about mind, and their knowledge of consciousness becomes explicit when they can describe how it is ontologically different from the way the natural world exists. That is, what they discover is *ontological naturalism*, defined here as the belief that the natural world is made of objects in space that exist independently of one another. That is relevant because knowing about consciousness in that way does not depend on the assumption that phenomenal properties are known by their immediate presence. It depends only on the unity of consciousness and how it can be the illusion described by the Cogito.

In other words, consciousness as the phenomenal intrinsic property of a bit of matter helping constitute the mammalian brain is not an efficient cause of anything that happens in the natural world—except for how the spatiotemporal structure of its coincidence with space enables reflective subjects to know that consciousness is ontologically different from the natural world.[24] Consciousness is just what it is like to be a mammalian

---

[24] If there were a spatio-material world in which matter had no proto-phenomenal intrinsic property, the metaphysical stage would unfold in the same way because the relevant bit of magnetic field matter in the mammalian brain would still have the kind

brain, and since it plays a strictly passive role in knowledge about it, it is fitting to call it a *pseudo-efficient cause.* Since modern intuitionistic metaphysicians do not recognize this, but rather, fall for the intuitionist illusion, they discover that they are conscious in the problematic form of mind-body dualism. They believe that intuitionistic Reason knows the Reality behind phenomenal appearances of all kinds, and since that Reality must be a substance with the unity of consciousness, they have to explain how it interacts with substances that are essentially divisible into parts that exist independently of one another. That is the belief in mind, and it is false. But since mind is conjured up by intuitionistic metaphysics, it is a *phantom Reality.* Thus, ontological science reduces mind to language-using mammals who are taken in by the illusion of intuitionism and believe that knowledge of phenomenal properties is caused by their immediate presence. Since this explanation of the identity of mind and brain is an implication of the explanation of Western civilization as a third stage of spiritual evolution, it is the discovery of a theoretical identification in ontological science in the same sense that the identity of water and H_2O was a theoretical identification discovered in physical science.[25]

This ontological explanation of the belief in mind explains why physical science cannot solve the hard problem of mind. The belief in mind becomes widespread in Western culture because mammals naturally believe that they

---

of spatiotemporal structure that explains the configuration of qualia in phenomenal space that is immediately present in our spatio-material world. Though there would be no vivid sensory qualia, it would still be like something to be the relevant bit of magnetic field matter in the mammalian brain because there would still be a phenomenal space whose parts are all immediately present. Reflective subjects would still refer to locations in phenomenal space and classify them by kinds in the same way. What they meant by *consciousness* would be different. But there would still be intuitionistic metaphysicians trying to show that Reason knows Reality behind Appearance.

[25] In both cases, it is the discovery of the identity of something found on one level of natural organization with something found on another level of natural organization. Water was something that is ordinarily recognized by properties that are relevant to reflective subjects leading lives on the macro-level in the natural world, and when physical scientists discovered that material bodies are composed of atoms on the micro-level, water turned out to be identical to collections of molecules called H_2O. Likewise, since the belief in mind is an inevitable development during the metaphysical stage, it is part of the evolution of a culture with the third level of linguistic organization, and thus, mind turns out to be identical to the brain because it is just a phantom substance caused by the intuitionist illusion in a metaphysician's brain.

know about phenomenal properties because of their immediate presence, and once they learn what the mind is, they cannot help believing that it exists. This goes for physicalists, and as ontological naturalists, they would like to believe that the mind is identical to the brain. But like Descartes, physicalists believe that the natural world is constituted by a single kind of substance, such a monistic ontology cannot explain how phenomenal properties are part of the natural world in a way that exposes the illusion inherent in consciousness. The only way that physicalists can explain how they know about phenomenal properties as something different from brain states is by assuming that knowledge of phenomenal properties is an effect of their immediate presence, and so they cannot even claim that the mind is identical to the brain without contradicting physicalism, much less show how physical science could discover the identity.[26]

(7) The Second Enlightenment

By tracing the origin of ontological science to intuitionistic metaphysics in Western civilization, ontological scientists will cut the Gordian knot of philosophy, and since that will bridge the gap between science and philosophy, it will unify them. Everything found by both the study of nature and the study of Appearance will be explained by the same *ontological causes*, that is, by space and matter as the first cause. Interactions of space and matter will explain the regularities described by the laws of physics, and since that implies that *physical and geometrical efficient causes* are jointly responsible for all the other regularities about change found in the natural world, ontological scientists will discover a series of evolutionary stages that lead eventually to the existence of reflective subjects. Knowing that the function responsible for the evolution of the mammalian brain is to serve as a faculty of imagination, they will discover that consciousness can be explained as what it is like to be a mammalian brain, and since the illusion of intuitionism will enable them to explain the role of consciousness in helping cause the metaphysical stage of evolution as a *pseudo-efficient cause*, they will discover an explanation of what holds necessarily in the natural world

[26] The inability of physical science to make such a discovery was the gist of the argument by which Kripke forced philosophers of mind to recognize many decades ago that they had not explained how mind-body identity could be an empirical discovery in science. Saul Kripke (1980) *Naming and Necessity*, Cambridge University Press, Cambridge.

that is complete—except for a regularity essential to choosing good goals that is discussed in the final chapter. (See Figure 11.)

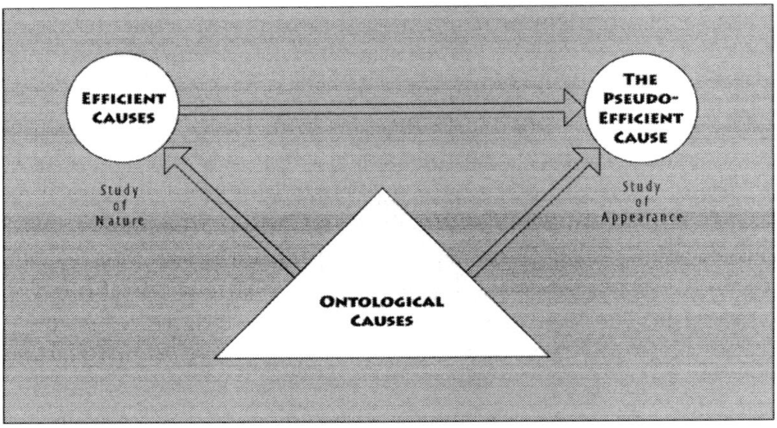

FIGURE 11 - CAUSAL COMPLETENESS OF SPATIO-MATERIALISM

As reflective subjects, ontological scientists will understand ontological explanations in naturalistic imagination (a cognitive faculty built into the brains of language-using mammals), and since they are conscious, they will be able to understand what intuitionistic metaphysicians meant by Reason knowing Reality behind Appearance. But to explain Western civilization as the metaphysical stage of evolution, they will have to *discount the illusion of intuitionism* in themselves. That is, ontological scientists will have to use their faculty of naturalistic imagination to explain their own reflection, and since understanding metaphysical arguments depends on seeing phenomenal properties as objects of intuition, they will also have to use their faculty of naturalistic imagination to explain their own consciousness as a pseudo-efficient cause. And when they discount the intuitionist illusion in this way, they will find themselves knowing Reality behind Appearance. They will realize that they have the cognitive power that metaphysicians call *Reason*. That is how science will become Reason. But since ontological scientists will know that their knowledge of Reality depends on the empirical method, rather than intuitionism, they will insist that their cognitive power is *naturalistic* Reason, or *natReason* for short.

The Second Enlightenment predicted by this trilogy is the dawning of the explanation of everything depicted in Figure 12. It is a revolution in culture that will begin with the discovery that space is a substance interacting with matter, and it will turn science into the perfect knowledge of Reality behind Appearance that has always been the goal of metaphysics. To be sure, it will not have the certainty claimed for intuitionistic metaphysics. Its immediate presence does not make it self-evident. But naturalistic metaphysics will still make good on the promise of metaphysics because instead of coming from a supposedly indubitable faculty of rational intuition, the certainty of naturalistic Reason will come from the completeness of its empirical explanation. It will explain every kind of regularity found in the natural world, and since nothing basic will be left unexplained, there will be no grounds on which anyone could reasonably doubt it. The completeness of naturalistic metaphysics will make it the perfect knowledge of Reality that philosophers have sought ever since ancient Greece, when the pre-Socratics and metaphysicians began the search.

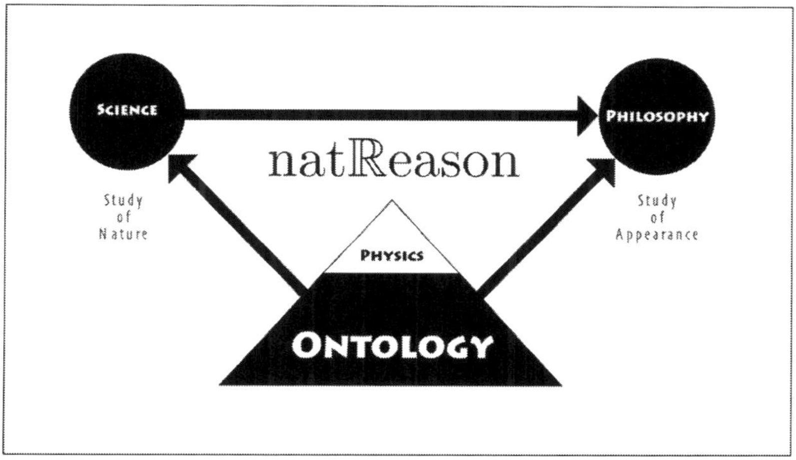

Figure 12 - natReason

When ontological science explains its own existence in this way, reflective subjects will have naturalistic Reason, and the natural world will already seem to be perfect because the existence of beings like them is necessarily part of it. Though the natural world is nothing but two kinds of substances that endure through time, their interactions generate regularities that imply that life evolves on suitable planets everywhere in an infinite universe. Life evolves by the reproductive mechanism, and it causes a series of inevitable stages that lead to the existence of reflective subjects who are eventually able to explain everything that happens necessarily in the natural world. Since they know that they exist necessarily in a world constituted by space and matter, they are necessarily parts of Reality that know the existence and nature of Reality. In effect, they are how the ontological causes of the natural world know that they cause it, and in a way, that seems complete. The effects of their interactions are so far-reaching that space and matter are arguably the kind of first cause that makes the most out of existence. This impression seems to be confirmed by the consciousness of reflective subjects who acquire this knowledge. Though their knowledge of being conscious involves discounting the illusion of intuitionism, they will see the phenomenal appearance of their complete understanding of the natural world as a vivid mirror of its perfection, and when they reflect on their aesthetic appreciation of its beauty and recognize that it is also part of the natural world that exists necessarily, they will see the perfection of Reality as even more complete than the perfection of ontological causes that merely bring beings into existence who understand how it makes their own existence necessary. Thus, the Second Enlightenment will make it hard for reflective subjects with naturalistic Reason to deny that the natural world is a perfect being.

But this perfection of theoretical Reason is only part of what ontological scientists will discover when they explain how the metaphysical stage of spiritual evolution leads to the perfection of practical Reason.

Chapter 4

Unification of Science and Philosophy (Volume III)

Practical Reason

Naturalistic Reason is knowledge of the True, and since the Good is part of the True, it is also knowledge of the Good. That enables Reason to serve its practical function. Reason is the new kind of BGS that evolves in spiritual organisms during the metaphysical stage, and since guiding behavior to goals that are good is the function for which it evolves, its practical function is more basic than its theoretical function. Historically, practical offspring of the 18th-Century Enlightenment are at least as important as science, its theoretical offspring. Besides political institutions that respect the moral autonomy of Rational beings, they include an economic system in which competition among capitalists seeking a profit increases everyone's affluence. Though both maximize the power of individuals, one does it by justifying political freedom and the other by justifying economic freedom.

With the completion of the First Enlightenment, the puzzles that perplex physics, specialized sciences, and philosophy will all be solved by the discovery that space is a substance that interacts with matter. But even if this cultural revolution is imminent, it will occur in the aftermath of the 18th-Century Enlightenment when metaphysical confidence in Reason has nearly petered out. Scientists will be disillusioned with theoretical Reason, and endarkenment will be spreading the relativism defended by intuitionistic philosophers who reject metaphysics to popular culture. Doubt about settling disagreements by arguing about them will be conventional wisdom, and basic rights, such as freedom of speech, will be eroding. Doubts about the ability of reflective subjects to take responsibility for what they do will lead political activists to assume that solutions to economic, social, and political problems all depend on increasing the power of experts in an administrative state to take responsibility for doing what is good for them. But physical science cannot tell experts what to do. It can predict the consequences of pursuing some goals, but scientists themselves warn that it cannot discover what ought to be done. Since those who deny that there is any such thing as the True also deny that there is any such thing as the Good, relativism about facts extends to values, and turning against their heritage, philosophers will spread the gospel of the futility of arguing about what is good. It will seem that, in the end, the sophists win. That might makes right.

At this point in history, almost the only remaining defenders of the practical offspring of the First Enlightenment will be those who believe in God. Since Christians and orthodox Jews believe that the difference between good and evil depends on God's purpose in creating the natural world, they believe that the difference is more than just an expression of sentiments, or something projected onto an indifferent natural world. But theists cannot prove the existence of a transcendent God, much less explain his purpose, so naturalists will insist that they admit that their belief in God is just a matter of faith and dismiss the belief in God as rationally indefensible.

However, the belief in God will be vindicated when the Second Enlightenment bridges the gap between science and religion, and this chapter describes what it implies about practical Reason. Though this argument is defended in the trilogy, the conclusion drawn here goes beyond *Volume III*, and since it is more speculative, it might be considered the proposal of another research project for scientists with naturalistic Reason. It shows that the evolutionary explanation of Western civilization implies that God is identical to the natural world, and like the discovery of the identity of mind and brain, it is a theoretical identification discovered by ontological science with the same status as the discovery of the theoretical identification of water and H_2O by physical science. On Earth, the identity of God and nature will be Christian pantheism. But since it is not clear how much of this identity is ontologically necessary, the final section speculates about some implications concerning what is good and the meaning of life.[27]

(1) The Nature of the Morally Good

After completing the First Enlightenment, scientists will trace the origin of ontological science to metaphysics in Western civilization, and they will discover how the perfect way of knowing to which philosophy aspired includes knowledge of the Good. Metaphysics began as Plato's response to

[27] For evidence of the intensity of interest in these issues, see the series of four debates between Jordan Peterson and Sam Harris about the nature of the good before large audiences in 2018. These debates about God are easily available on the Internet, and more are now promised.

Socrates, who taught Athenians that *to know the good is to do the good.* Though Socrates did not know the nature of the good, he was looking for knowledge of it that would cause Rational beings to choose the good. He refuted the belief that the good is whatever the gods love by asking, Is the good good because the gods love it, or do they love it because it is good? The questions was rhetorical. Goodness had to be a property of what was chosen. But Socrates famously found no one in Athens who could describe a property that showed that those who knew it would do what is good, Athenians sentenced him to death for doing philosophy in this way, that is, for his love of practical wisdom. Plato's goal as a metaphysician was to answer Socrates' challenge. To show how this was possible, Plato argued that Reason can know the True, and since the Good was part of the True, he claimed that knowledge of the True would cause Rational beings to do what is good. But ontologists will see the metaphysical arguments exchanged in ancient Greece as failing to pass the Socratic test of practical wisdom.

Plato held that Reason knew the Form of the Good, and since the Good Itself was the origin of all the other Forms, it revealed what was good for the visible objects that imitated them in the realm of Becoming. But to explain why Rational beings chose to do what conformed to the Good, all Plato had to offer was an analogy to erotic love. He described the Good Itself in the realm of Being as arousing Platonic love for what is good in the realm of Becoming. But that was just to assert that knowledge of the Good causes Rational beings to choose what is good.

Aristotle explained the Good as the *final cause*, or the end for the sake of which *natural change* occurred. The final cause was the *actualization* of an essential form that exists *potentially* in each particular thing. But this regularity about change was the only reason he offered for believing that the final cause was good. To be sure, teleological causes are different from efficient causes. They attribute change to a future state rather than a past state. But there is no such cause in the natural world, and even if there were, final causation would be just a different kind of regularity. Since knowledge about what happens regularly does not imply that it ought to happen, Aristotelian teleology does not pass the Socratic test of practical wisdom. Since Rational animals could know their final cause, it could determine their behavior. Knowing that virtue is choosing the mean between extremes, for example, courage between cowardice and foolhardiness, they could choose to do what is good. Indeed, Aristotle held that Rational animals are responsible for their virtue because virtue is a habit of choosing the mean between extremes that they can acquire by choosing the mean in situations where it is easy. But that shows only how knowledge of their final cause could cause Rational animals to be virtuous, if they chose to, not that they ought to choose to be virtuous.

Furthermore, it is noteworthy that practicing choosing does not require a spiritual desire any more than practicing the use of a sword requires fighting. Ontological scientists will know that, as members of spiritual organisms, being moral is the good on which the goodness of other goals depend, and it clearly depends on having a spiritual desire. They will notice that the primacy of moral goodness plays no part in the final cause of the city-state. In Aristotle's view, its final cause was little more than how virtuous Rational beings who are friends manage to live together. And though, on the other hand, the spiritual desire plays a central role in Plato's view of the ideal city-state, ontological scientists will see in his collectivist view of our spiritual nature a dangerous error. Believing that only some reflective subjects had the cognitive power to know the Forms, Plato made philosophers kings, and they would use auxiliaries, that is, disciplined police forces, to make sure that all the other members of the *Republic* did what only philosophers could know was good for them.

Ontological scientists will know that what makes the morally good good for reflective subjects is the spiritual nature of their form of life because they will know that the nature of life explains the nature of the good. Life is a product of the reproductive mechanism, and as we have seen, this origin implies that a form of life begins with the evolution of a new kind of BGS. Since its function is to guide behavior, it is a *choosing machine*, and since choices matter, the nature of life entails a difference between good and bad. In spiritual organisms, the function of its BGS is to guide the behavior of its members to goals that are good because they control conditions that make its reproduction possible. The behavior of the members controls such conditions on both the social and the individual levels, and when the members are reflective subjects, cultural evolution by pragmatic selection makes moral rules a basic part of the shared cultural plan that provides spiritual input to them. They are rules that forbid doing harm to other members unnecessarily and require them to tell the truth, keep their promises, help others in need, and the like. Morality is the balance that cultural evolution reaches between goals that are good for individuals and goals that are good for the group as a whole, that is, when cooperation in pursuit of common goals is most powerful. As a basic part of their cultural plan, morality has spiritual authority, and reflective subjects are able to do what authoritative spiritual input prescribes because they have a spiritual desire (derived from the dominance hierarchy of social animals) that can override other desires. Their will is their representation of what they choose, and it is free because their neural BGS is a geometrical cause with the function of choosing. That means that, like all other living organisms, they could always have chosen otherwise. Reflective subjects are different only because they can represent the causes of their behavior as part of the process of choosing. When they decide whether to do what spiritual input prescribes, they know that they are able to choose otherwise because they have a spiritual desire. Thus, they can take responsibility for their choices. They have a free will. They can do what morality requires by choosing to act on their spiritual desire.

Ontological scientists will know that it is good for reflective subjects to prefer the morally good over contrary desires and good for them not to act in their self-interest when it is morally wrong. That is how they must choose to lead lives as parts of spiritual organisms. Athenian metaphysicians did not recognize that morality is the paramount good for each and every member of a spiritual organism. But a marriage of a religion with Platonic metaphysics during the ancient era laid the foundation for a culture in the modern era that did recognize it. Religion could use Reason to know the existence of a God who created the natural world in the way that Plato had used Reason to know Forms in a realm of Being beyond visible objects in a realm of Becoming. And Platonic metaphysics acquired an explanation of why the morally good is good that could pass the Socratic test of practical wisdom because unlike the Form of the Good, God was a Rational being who created the natural world for a purpose.

Knowing that what makes the morally good good for reflective subjects is their spiritual nature, ontological scientists will see an outline of their own explanation of morality in Christianity. It holds that God had a purpose in creating the natural world for finite beings created in his own image, and according to the gospel, God sacrificed his own Son to forgive their sins and make it possible for them to live together in a heaven on Earth by choosing good over evil. The gospel is a kind of knowledge about the nature of the morally good that can cause those who believe it to choose to do what is right, so it passes the Socratic test of practical wisdom. It recognizes that doing what is good requires resisting the temptation to sin. The need for a spiritual desire to resist doing what is evil is the more profound practical wisdom that Athenian metaphysics acquired from its marriage with Christianity, and after the fall of Rome, this super-naturalistic view of the spiritual nature of reflective subjects became the foundation of the culture in which modern metaphysics evolved.[28] The Christian explanation of how the Good is part of the True was close enough to the moral rules that maximize the power of cooperation in psychological-level spiritual organisms for the constraint that it imposes on conclusions about which goals to pursue after the fall of Rome to make an essential contribution to the practical offspring of the Enlightenment.

Christianity fit neatly together with Platonic metaphysics because the belief that God created the world was an explanation of the relation between Being and Becoming. This was a way of defending the belief that there is only one God, and though monotheism originated in Judaism, the marriage was with Christianity. It had split off from Judaism, and its marriage with Plato's metaphysics turned the Form of the Good into a perfect Rational being outside space and time who created the natural world for a purpose. The purpose was to provide a place for finite beings fashioned in his own image to lead lives making choices about how to live. The Christian commandment was to love God and to love one's neighbor like oneself, and since it was an idealized description of moral goodness, Christianity was a recognizable image of the spiritual nature of reflective subjects entailed by the ontological explanation of their evolution. But the Christian explanation was otherworldly. Unlike all the other animals on Earth, human beings were created in the image of God, and since God willed the natural world into existence freely, human beings also had a free will. God created the natural world for a purpose, and Christians believed that, by sacrificing his son to forgive the original sin of disobeying his command in the Garden of Eden, God was promising that, if human beings believed in Christ and chose freely to follow the Christian commandment, they would be rewarded by living together with one another in heaven instead of being damned for eternity. Christ forgave even those who crucified him, and since he was the ideal to which God meant Christians to aspire, the goodness of morality was clearly greater than self-interest. Since God created all human beings in his own image, they were all morally equal, like members of a single spiritual organism who can see into one another's minds. Differences between gentiles and Jews, women and men, blacks and whites, and the like were not morally relevant. Even slaves and masters were equal, since Christ himself died the death of a slave. It was good for everyone to choose good over evil

---

[28] For an account of the difference Christianity made, see Tom Holland (2019) *Dominion: The Making of the Western Mind* (U.S. edition subtitled 'How the Christian Revolution Remade the World'), London: Little, Brown.

because resisting the temptation to sin against one another was how God's creatures could live together in a heaven on Earth. Knowledge of this explanation of the Good gave Christians a reason for being moral. To be sure, they did not know God's purpose in creating the natural world and populating it with beings whose free choices would determine their fates, especially, if it were the everlasting fates of immortal souls. But since human beings were created in God's image, they knew they had the kind of cognitive power that would enable them to understand God's purpose, even if they could not know his purpose until they were in the presence of God in heaven. According to the book of *Revelations*, that would happen when Christ returned, their bodies were resurrected, their lives judged, and their eternal fates determined. That was not part of the gospel. But it was a promise of a perfect outcome of life on Earth, and it prepared the way for popular acceptance of the Enlightenment belief in cultural progress on Earth.

With the rise of Christianity in the Roman Empire, the gospel spread to everyone, and this heritage from the ancient era gave language-using mammals in the West a second chance during and after the so-called *Dark Ages* to argue that Reason knows Reality behind Appearance. A God that creates a natural world for a purpose is Rational, and since God created all human beings in his own image, Christianity spread the belief that everyone is Rational. That is what made them all morally equal. Thus, from the beginning of the modern era, Western culture was founded on metaphysical religion, and since it assumed that everyone was capable of a perfect way of knowing and acting that passed the Socratic test of practical Reason, ontological scientists will see Christianity as helping cause the cultural evolution that became the Enlightenment in the 18th-Century that respected the moral autonomy of finite rational beings by giving them political and economic freedom.

(2) Practical Offspring of the First Enlightenment

The liberal political institutions spawned by the Enlightenment were mainly due to Locke's use of the social contract to protect against tyranny by showing the limits of legitimate government. He used the social contract to represent the source of the spiritual authority of political power. An agreement among Rational beings is what created government, and thus, its use of violence to enforce its laws was justified only insofar as it served the functions for which government was created, such as protecting the rights of citizens and punishing those who commit crimes. But since the social contract created government in a state of nature, the Rational beings who agreed to it had to be moral in the state of nature, and Locke held that they could all be trusted to act morally because they knew that they were all created by a transcendent God whose purpose required them to choose good over evil.

Locke's argument was explicitly founded on the dualism of God and nature inherited from ancient metaphysics. But with the rise of science, supernaturalism lost favor, and naturalists also needed an explanation of morality.

Hobbes was a materialist who used the social contract to construct what he called the *Leviathan*. Though he did not deny the existence of God, his argument was not based on a theistic explanation of the spiritual nature of reflective subjects. He assumed that beings like us were by nature rationally self-interested, and thus, he saw the state of nature as a war of all against all in which the *life of man [was] solitary, poor, nasty, brutish, and short*. The only way to leave the state of nature was to establish a sovereign power whose commands would keep peace among them. Since the sovereign would still be in the state of nature, he would act in his own self-interest, and so he could be trusted to make commands that keep the peace. His authority would be respected because he would have the use of violence to enforce his commands. With everyone in such an autocratic state acting in their own rational self-interest, the Leviathan was a naturalistic explanation of morality that did not distinguish it from political authority. But that was not what metaphysicians were looking for. They expected knowledge of the Good as part of the True, rather than self-interest, to be what causes Rational beings to be moral. And ontological scientists will see in Hobbes' sovereign the spiritual desire writ large in much the same way that it was writ large in Plato's defense of collectivism in the *Republic*.

Rousseau was also a naturalist who used the social contract to explain moral goodness. Though he believed in God, he was a deist. He believed that since God had created the natural world with an order that would fulfill his purpose without any further intervention, Rational beings could learn the difference between good and evil from nature. And the Enlightenment belief in cultural progress suggested that morality arose from more basic aspects of the natural order. Rousseau used the social contract to defend what he called the *general will*, and though it is commonly denigrated as defending the kind of collectivism that inspired the excesses of the French revolution, ontological scientists will recognize that the content of the general will was the moral balance that reflective subjects must will to be members of a spiritual organism. Ontological scientists will see in Rousseau's social contract an early attempt to explain the origin of morality as a necessary part of the evolution of spiritual organisms. As he describes the social contract, it is an agreement that gradually evolves as compassionate language-using animals learn the rules that they must follow in order to cooperate in the pursuit of shared goals.

Kant also defended a contractarian explanation of morality in a way that did not depend on supernaturalism. His constructivist theory of Reason implied that God is a thing in itself that cannot be known, if he exists at all. But it also implied that morality was a necessary aspect of the phenomenal world. Kant held that Reason serving its practical function entailed (synthetic a priori) knowledge of morality in the same way that Reason serving its theoretical function entailed (synthetic a priori) knowledge of mathematics and laws of physics. Whatever a Rational being willed, the practical function required that it be possible for Reason to will that it become a universal law of nature. The *categorical imperative*, as it is called, entails the moral equality of Rational beings, and since that determines the content of morality, it justified the respect for the moral autonomy of Rational beings on which the social contract founds liberal political institutions.

Social contract theories all explained the goodness of morality and political institutions by the nature of individual human beings, and except for Hobbes, the spiritual nature they described is recognizably an inheritance from Christianity. Ontological scientists will see this development as an expression of metaphysical confidence in Reason that made Rational subjects as powerful as possible. It gave them as much political freedom as is compatible with everyone having equal rights, and it gave them as much freedom in providing the bodily means required to exercise political rights as possible. This maximized the power of both the individuals and their spiritual organisms.

In the Middle Ages, legal institutions that protected the rights of individuals were extended to associations of individuals. Confidence in Reason enabled the Church to protect itself from monarchical power by restoring Roman law and acquiring authority over its own property and such affairs, and the extension of these legal rights to other associations, such as universities and cities, were the foundation for voluntary economic associations that replaced class structures based on birth that had enabled civilization to evolve. Capitalists established associations by investing money and hiring workers to carry out production. They expected to earn a profit, and since they had to compete with one another in selling commodities that consumers were willing and able to buy, they used new technologies spawned by advances in science to produce a growing variety of commodities ever more efficiently. Historically, this development is known as the *Industrial Revolution*. But it is capitalism, and ontological scientists will see this economic system as an embodiment of the reproductive mechanism within spiritual organisms. Capitalist associations are constituted by the legal rights that government grants to associations, and since they are double-action geometrical causes that go through annual cycles of reproduction in a region with finite resources, production evolves by capitalist selection within spiritual organisms. Since the means of survival are produced in a way that becomes more powerful at providing commodities that consumers are willing to buy, it increases the power of the spiritual organism in which it is embedded. But capitalism also increases the power of its members because a class structure in which membership depends on the accident of birth is replaced by associations that depend on individual choices in free markets.

Liberal political institutions and capitalism were the practical offspring of the 18th-Century Enlightenment, complementing science, its theoretical offspring. But its practical offspring maximized the equal power of the finite Rational beings, and since Christianity held that they were all created in God's image, ontological scientists will recognize them as a legacy of the marriage of Christianity with Platonic metaphysics. Christian dualism explained the natural world as just part of Reality, and when representative realism about perception replaced direct realism on Earth, a basically different kind of metaphysical argument was possible. It enabled Descartes to show that reflective subjects are conscious, and since he discovered it in the problematic form of the ontological difference between the unity of consciousness and the unity of the natural world, he turned up the second basic ontological problem caused by intuitionistic metaphysics. In the absence of a solution to the mind-body problem that could pass the pre-Socratic test of ontological coherence, metaphysics became increasingly unbelievable during the modern era, and as we have seen, in the aftermath of the 18th-Century Enlightenment, an abyss opened up between science and philosophy. After the incomplete scientific rejection of intuitionistic metaphysics, naturalists became increasingly disillusioned with Reason, and intuitionistic philosophers unbridled by the empirical method, spread doubts about the existence of the True and the Good to popular culture. With postmodern philosophers preaching the gospel of relativism, the tide of endarkenment rose.

The previous chapter summarized the argument of *Volume III* that predicts the unification of science and philosophy. The ontological explanation of consciousness enables ontological scientists to explain Western civilization as the metaphysical stage of evolution. Recognizing the illusion inherent in consciousness, they will solve the mind-body problem, cut the Gordian knot of philosophy, turn ontological science into naturalistic Reason, and enable Reason to serve its theoretical function in the most complete way possible. But it remains to be shown how Reason will also serve its practical function in the most complete way possible.

The ontological problem that confronted Descartes at the beginning of the modern era was not just the dualism of mind and body in the natural world but also the dualism of God and nature inherited from ancient metaphysics. Since mind-body dualism was a problem in the world created by God for finite beings fashioned on his own Rational nature, dualism was compounded. But ontological scientists will discover similar solutions to both ontological problems. As the mind turns out to be identical to the brain, so God will turn out to be identical to the natural world. Though physical scientists now recognize that the belief in mind poses a problem that they must solve, they cannot solve it, so they will be surprised to learn how science discovers that the mind is identical to the brain. But since they do not even recognize that the belief in God is a problem, they will be surprised to learn that God is identical to the natural world. The discovery of this identity will complete the Second Enlightenment because as the solution to the mind-body problem leads to the perfection of theoretical Reason, so the solution to the God-nature problem leads to the perfection of practical Reason.

(3) Theoretical Identification of God and Nature

Since naturalists assume that the natural world is what exists, they expect science to discover that the mind is identical to the brain. But it is almost inconceivable that the Christian God is identical to the natural world because a God who created the natural world for a purpose is a Rational being, and a world of objects that exist outside one another in space cannot be a person. Naturalists who believe in God are pantheists, and since physicalists cannot explain moral goodness by God's purpose in creating the natural world, they must give up the Christian explanation of how knowledge of the Good causes Rational beings to choose to be moral. This loss was acceptable to Spinoza, the modern metaphysician who defended pantheism, because as a Jew, he believed that Rational beings could be moral without accepting the Christian gospel. Indeed, he denied that the will was free and reduced morality to rational self-interest. And since the rejection of supernaturalism denied the existence of a transcendent God, naturalists typically explained belief in God as a projection of our human nature onto a perfect being outside space and time, and scientists did not take a position on whether we ought to be moral.

When science based on ontology discovers that Western civilization is the metaphysical stage of spiritual evolution, it will discover that God is identical to the natural world. But this discovery depends on discounting another illusion that otherwise misleads beings like us. As we have seen, the discovery of the identity of mind and brain (like the discovery of the identity of Platonic Forms and meanings of general terms constructed in mammalian linguistic faculties of imagination) depends on discounting the illusion of intuitionism. But the discovery of the identity of God and nature will depend in a similar way of discounting the *illusion of purposiveness*, the belief that the goodness of goals causes purposive behavior. Purposes seem so obviously explanatory in this way that the explanation above about what enabled Athenian metaphysical arguments to pass the Socratic test of practical wisdom took this illusion for granted. Indeed, the illusion of purposiveness is what Socrates had in mind when he set the standard for practical wisdom by asking rhetorically, *Is the pious pious because the gods love it, or do the gods love it because it is pious?* The latter alternative implies that the goodness of what is chosen is what causes one to choose it.

The illusion of purposiveness is basically different from the illusion of intuitionism. The intuitionist illusion depends on an illusion inherent in *consciousness*, where consciousness is what it is like to be a mammal. Since consciousness described as something distinct from the mammalian brain is not a cause of anything that happens in the natural world, it plays the role of a pseudo-efficient cause in helping cause the metaphysical stage of evolution. Consciousness is merely what it is like for language-using mammals to participate in the process. But the illusion of purposiveness is inherent in the nature of *reflection*, and since reflection is a power of the geometrical cause of the behavior of reflective subjects, it is an efficient cause. There is, nonetheless, an illusion inherent in reflection. Mammals whose BGS has the use of psychological sentences naturally assume that they choose goals because they are good, and the belief in such *purposes* is false because the goodness of goals is not what causes the behavior aimed at them.

The illusion of purposiveness plays a different role from the illusion of intuitionism in the explanation of Western civilization as the metaphysical stage because it is an additional illusion that ontological scientists will use to explain why ancient intuitionistic metaphysics passes the Socratic test of practical wisdom. Athenian metaphysics had a superficial way of passing the Socratic test of practical wisdom because it did not explain moral goodness, and as we have seen, the marriage of Platonic metaphysics with Christianity is what enabled ancient metaphysics to lay the foundation for a culture to evolve after the fall of Rome that could evolve by Rational selection into modern intuitionistic metaphysics. It explained why Rational beings ought to be moral even when it is contrary to self-interest or opposed by strong bodily desires as what God intended for them in creating the natural world for them. They would lead lives choosing freely between good and evil with the eternal fates of their soul at stake. Since God had a purpose in creating the natural world, he must have chosen this goal because it was good, and belief in God promised, at least, an explanation of why it was good for Rational beings to resist the temptation to sin and love their neighbors like themselves that they could understand.

Purposiveness is an illusion inherent in reflection because mammals with the use of psychological sentences can represent the causes of their behavior in the very process of guiding it. Reflective subjects represent their choice of which goal to pursue in the situation as their *intention* or *will*, and they explain their intention by identifying its purpose. That is, they explain their behavior by the goodness of the goal at which it aims because what matters to choosing machines is choosing goals that are good. But reflective subjects do not necessarily know the nature of the good. Though they know that their desires help cause their behavior, desires do not seem to be what determines the goals they pursue because they know that their intentions are the result of choosing among goals in the situation. They consider different plans, and since each may be a different combination of means and ends, the choice depends on comparing plans in imagination and choosing among them. They explain their purpose in the situation by the goodness of the goal they choose. But when asked to explain their purpose in choosing that good goal, they point to the goodness of a further goal. When the goodness of goals is explained as instrumental in a further goal, a regress begins that points to an ultimate goal, and since the regress is endless, they are never able to define the ultimate goal adequately. Hence, they usually just stop and insist that some goal is intrinsically good. That is what Plato did when he claimed that the Form of the Good is what makes all the other Forms in the realm of Being good and, then, used their goodness to explain what is good for visible objects in the realm of Becoming imitating them, including the Rational beings who know about them. The illusion of purposiveness is what made Platonic love of Forms a plausible explanation of why Rational beings do what is good.

The nature of what is good is elusive because the choice of a goal is not caused by its goodness. There is no ultimate purpose. Aristotelian teleology was just the use of the illusion of purposiveness to explain what happens for the most part in the natural world, and final causes were abandoned in the early modern era by natural philosophers looking for efficient causes. Though the purposive appearance of living organisms is undeniable, physical scientists were not taken in by it because they could explain purposiveness as an appearance. Ever since Darwin's explanation of the origin of species by natural selection, physicalists have understood how purposiveness could be explained by efficient causes. While final causes depend on regularities about change in which a future state seems to cause what happens at present, Darwin's combination of random variation and natural selection replaced them with causes by which the past determines what happens at present. To say that a trait is *functional* is to say that it has a certain effect and that it exists because it has that effect.[29] The Darwinian explanation of functional traits showed that purposiveness was just an appearance of efficient causes. But physical scientists will be surprised to learn that Darwinism overlooks a deeper error in the illusion of purposiveness, one that omits the role of goodness.

Since Darwinism does not entail progressive evolution, it leaves physical scientists with an explanation of the appearance of purposiveness in which the goals at which behavior is aimed are accidental and not necessarily good. That is, goodness is also just an appearance. But ontological scientists will have a deeper explanation of what is erroneous about final causes because they explain evolution by the reproductive mechanism. Instead of explaining goodness away, they will discover that goodness is an essential aspect of the nature of life. Goodness comes into existence with life because life begins with the evolution of a BGS, and its function is to choose which goal among various goals to pursue in each situation as it arises. The goals that a BGS normally chooses are good, and the difference between good and bad matters to living organisms because their form of life depends on a BGS choosing goals that are good. What is illusory about purposiveness is not that goodness is a mere appearance. The

---

[29] One of the generally accepted conclusions in analytic philosophy is this explanation of *functions*. See Larry Wright (1973), "Functions," *The Philosophical Review*, Vol. 82, No. 2, pp. 139-168.

illusion is that talk of purposes gets the causal order between goodness and choosing backwards. What is illusory about purposiveness is the belief that the goodness of the goal causes the choice of the goal. That is backwards because the goodness depends on the choosing. The good goal is the one that the BGS normally chooses. And since the goal really is good for the living organism, goodness is not just an appearance. Thus, when ontological scientists discount the illusion of purposiveness, they will not deny the reality of goodness but, rather, recognize that the causal relation between choosing the goal and the goodness of the goal is just opposite to how it appears. Their explanation of this causal relation is what enables ontological scientists to discover that God is identical to the natural world, while physical scientists cannot.

It may seem that goals are good because they help control conditions that make it possible to reproduce if and when the time comes. But that is to confuse *what is good* with *what is functional*. The BGS is a product of the reproductive mechanism, and it is functional because it has the power to choose goals that control relevant conditions. As a result of progressive evolution, the goals that a BGS chooses are increasingly able to control conditions that enable it to reproduce when (or if) the time comes. It acquires new functional powers. But many geometrical causes that evolve by the reproductive mechanism that have no BGS are functional in that sense. They include not only arguments exchanged in spiritual organisms and capitalist associations but also the double-action geometrical causes whose progressive evolution eventually gives them a BGS and turns them into triple-action geometrical causes, that is, living organisms, the kind of geometrical cause that causes its own gradual change in the direction of maximum functional power. To be sure, since the function of the BGS is to guide behavior, the goals it chooses tend to promote its reproduction. But controlling relevant conditions is not what makes the goals good. Functionality is just a property of goals that gradually increase during a stage of evolution. What makes goals good is that they are normally chosen by a BGS. Choosing those goals is good for a living organism because choosing is what matters to a choosing machine. Forms of life are defined by their BGSs, and since goodness is a property that depends on the nature of life, goodness comes into existence when a form of life evolves.

If Athenians knew that goodness is not a cause of behavior, they would not have believed that Plato's metaphysics passed the Socratic test of practical wisdom. They would have known that the Form of the Good could not make all the Forms that follow from it good and, thereby, explain what is good for the visible objects in the realm of Becoming that imitate them. But as we have seen, even though the illusion of purposiveness made it seem that Platonic metaphysics could pass the Socratic test of practical wisdom, it was a superficial explanation of the nature of the good because it did not explain moral goodness. Though acting justly was a virtue, it was just one virtue among many that a Rational being might have. The failure to recognize the primacy of moral goodness might be expected of intuitionists who believed that they could know Reality behind Appearance because Reason is a cognitive power of individuals. Morality has to do with the rules governing interactions of reflective subjects in spiritual organisms, and to lay the foundation for the evolution of modern intuitionistic metaphysics by Rational selection after the fall of Rome, what ancient intuitionistic metaphysics needed was a more profound way of passing the Socratic test of practical wisdom, one that explained why Rational beings ought to choose to be moral.

As we have seen, what made this possible was the marriage of Platonic metaphysics and Christianity. Christians had to be taken by the illusion of intuitionism and believe that Reason enabled them to know the existence of the realm of Being behind the realm of Becoming. But since they already believed in a transcendent God, they described Plato's phantom Reality as a Rational being who created the natural world for a purpose. Since all religions explain why reflective subjects cooperate in the pursuit of shared goals, they all explain why it is good to follow moral rules. On Earth, Platonic metaphysics was married to Christianity, and as that religion explained the meaning of life, moral goodness was central. Life was the struggle to resist the temptation to sin and do what morality requires. For Christians, knowledge of God was knowledge of the True, and that entailed knowledge of the Good because God created the natural world for a purpose. He willed the existence of a world in which finite Rational beings fashioned in his image led such lives because, for some reason, it was good. Even though they could not know how God's goal made morality the primary good until they were in God's presence in heaven, they could believe that there was an explanation of why morality is good that they could understand that would cause them to choose it. Hence, Christian metaphysical theology passed the Socratic test with a more profound practical wisdom than Platonic metaphysics.

Like Plato's explanation of the nature of what is good, however, ancient metaphysics depended on the illusion of purposiveness. The goodness of a goal is not the cause of choosing it, and if Christians knew that, the belief in a transcendent being who willed the natural world into existence would not have caused them to do what he commanded. The creator might have been a masochist who created finite beings in his own image to make them suffer. He might have been a sadist or evil in some other way. And by giving them immortal souls, this evil transcendent being might have intended to force them to obey his commands by punishing disobedience with an eternal afterlife in a much worse hell. In other words, without the illusion of purposiveness, no religion would enable ancient metaphysics to explain why we ought to be moral in a way that passed the Socratic test of practical wisdom.

However, if Christians discounted the illusion of purposiveness in the way that ontological scientists do, they would conclude that the God in which they believed is identical to the natural world. That is, they would know that the natural world is constituted by space and matter, and as we have seen, that would enable them to discover the illusion inherent in consciousness. Hence, they would discount the illusion of intuitionism and recognize that Plato's realm of Being beyond Becoming is a phantom Reality. As naturalists, they would know that what exists is the natural world, and since they would understand how the illusion of purposiveness is inherent in reflection, they would not deny the existence of God, as physicalists do. Instead, they would recognize that the natural world plays the role of God in the nature of the good in a way that causes Rational beings to choose to be moral. To be sure, spatio-materialism would explain the Good as part of the True in a different way from Christian theism. But ontological causes, as the first cause, would justify the primacy of moral goodness in a way that is just as basic as the belief that God created the natural world. Interactions of space and matter cause life to evolve on all suitable planets by a series of inevitable stages that leads to the existence of conscious reflective subjects, and since they are members of spiritual organisms, the nature of their form of life entails an explanation of the Good that shows why Rational beings ought to choose to be moral even when it is contrary to their self-interest or it is opposed by strong animal desires. Like Christian theism, Christian pantheism would show that the difference between good and evil is essential to the nature of Reality.

To be sure, Christians who discount the illusion of purposiveness will explain their own behavior differently. They will not believe that the goodness of some goal is why they ought to be moral. They will know that what makes morality good is that their BGS normally chooses good over evil. Though they may still describe their behavior has having a purpose, they will discount the illusion of purposiveness and recognize that its goal is good because of how it is caused by their beliefs and desires. That is how their neural BGS chooses when it is functioning properly. Since reflective subjects can represent the causes of their behavior in the very process of causing it, they will use their spiritual desire to do what morality requires when they are tempted to do otherwise, and they will know that that is good for them because choosing to obey moral rules is what matters to reflective subjects. Goodness is entailed by the nature of life, and morality is entailed by the spiritual nature of their form of life because rational beings must choose to be moral to live lives as members of a spiritual organism. Christians who acquire naturalistic Reason will be pantheists, and knowledge of the Good will cause them to be moral because they will know that being moral is a condition of the goodness of other goals, that is, as the condition of living the life of a Rational being. That explanation of moral goodness will pass the Socratic test of practical wisdom, and it will pass with a higher grade than Christian theism because it will be a more complete explanation of moral goodness. It will not be founded on the promise of an explanation that they will not know until they are dead and in the presence of God. Christians who have naturalistic Reason will be in God's presence as they live on Earth.

In sum, ontological scientists will discover that the Christian God is identical to the natural world. They will predict that when ontological science becomes naturalistic Reason, Christians will recognize that what exists most basically in the natural world gives them everything they really wanted from Christian theism. Like Christian theism, Christian pantheism explains the meaning of life by the priority of moral goodness entailed by the spiritual nature of Rational beings. The difference is that their spiritual nature is this-worldly, rather than other-worldly. Furthermore, the choices that they make will have everlasting significance because they contribute to the life of a spiritual organism that is potentially immortal. That kind of immortality is also this-worldly. But it is still personal immortality because it depends on what individual Rational beings do. To be sure, only what Christians call the *soul* is immortal. The body and its consciousness must die. But as ontological scientists will predict, when Christians think about how boring consciousness would be outside space and time where nothing happens, they will recognize that, if immortal consciousness were possible, it is not something they really want. They will accept life on Earth as part of a spiritual organism for a finite time as a gift for which they feel profoundly grateful.

Such an ontological explanation of how naturalistic Reason serves its practical function will reveal that God is identical to the natural world in much the same way as the ontological explanation of how naturalistic

Reason serves its theoretical function will reveal that the mind is identical to the brain. The belief in mind turns out to be a belief about the brain that is false because reflective subjects are taken in by the illusion of intuitionism and believe that the immediate presence of their phenomenal properties is what causes their knowledge of them. That is, the mind is a phantom Reality, and since the marriage of Platonic metaphysics and Christianity promoted the Form of the Good to a God who created the natural world, Christians will recognize that the God in which they believed is another phantom Reality. Their belief in a transcendent God will turn out to be a belief about the natural world that is false because reflective subjects are also taken in by the illusion of purposiveness and believe that goodness is the cause of behavior aimed at goals. Since the discovery that God is identical to the natural world will be an implication of the explanation of Western civilization as the metaphysical stage of evolution, it will be the discovery of this identity in the same sense as their discovery of the identity of mind and brain and the discovery of physical science of the identity of water and H^2O.

(4) The Perfection of Practical Reason

The identity of God and nature will be a more surprising discovery than the identity of mind and brain, and what it implies about the perfection of practical Reason will be more speculative than what it implies about the perfection of theoretical Reason.

When ontologists discover that space is a substance that interacts with matter, science and philosophy will be divided by an abyss. While physical science studies the brain as part of nature, philosophy studies mind as part of Appearance. When ontological scientists explain Western civilization as the metaphysical stage of spiritual evolution, they will discover that the mind is identical to the brain, and that will cut the Gordian Knot of philosophical puzzles. This unification of science and philosophy turns science into naturalistic Reason and shows how it serves its theoretical function perfectly. But the denial of supernaturalism was an exceptional point of agreement between scientists and philosophers in the aftermath of the Enlightenment, so naturalists and intuitionists will both will be surprised to learn that naturalistic Reason also discovers the identity of God and the natural world. Indeed, Christians and Orthodox Jews will be the only remaining defenders of the practical offspring of the 18th-Century Enlightenment, and all those who believe in a transcendent God will be required to admit that their belief is a matter of faith. The surprise will be that naturalistic Reason proves the existence of God.

In the preceding section, the explanation of Western civilization as the metaphysical stage of evolution assumed that Christianity was the marriage partner of Platonic metaphysics, helping the ancient era lay the foundation for culture to evolve into modern intuitionistic metaphysics by Rational selection. It explained goodness in a way that caused believers in God to choose to be moral even when it was contrary to self-interest and bodily desires. But Christianity is a particular historical religion, and on other planets, Platonic metaphysics might be married to other kinds of religions. Overcoming the moral deficiency of Platonic metaphysics is part of the metaphysical stage, since it is part of a regularity about the overall course of evolution that must hold on all suitable planets, and ontological scientists will want to know how much of this development in Western civilization was ontologically necessary and how much was contingent.

All religions have an answer to the question, *Why be moral?* because that is a function of religion as an essential part of the culture of psychological-level spiritual organisms. When language-using mammals evolve the use of psychological sentences, they find themselves cooperating in the pursuit of shared goals, and they expect an explanation of why they are so different from other animals. Since reflective subjects explain their behavior by purposes, the answer that evolves in culture is a religion that explains the purpose of their cooperation. Since there is no such thing as an ultimate goal, religion is just a myth about gods or, at least, something mysterious, called the holy, that is so good in itself that it is worthy of worship and sacrifice. However, since cooperation depends on rules governing how members treat one another, moral rules evolve as part of religion, and all religions offer explanations of why reflective subjects ought to be moral.

On Earth, after the fall of Rome, Christianity became the foundation of Western culture and led to the practical offspring of the 18th-Century Enlightenment. But ontological scientists will want to know how much of Christianity is necessary on all planets where life evolves. Contingent conditions often play a significant role in determining the course of evolution. For example, the early impact of an asteroid gave Earth a moon that caused days to be much shorter at the beginning than now, and since that speeded up the reproductive cycles of RNA, it must have shortened the time required for reflective subjects to evolve. And the progress of evolution may have been slowed down by the age of dinosaurs since nearly 200 million years had passed before another asteroid gave mammals a chance to prove their inherently greater power by competing with dinosaurs on a more even playing field. It may even be possible that the practical function of Reason that led to the practical offspring of the 18th-Century Enlightenment on Earth does not evolve on some planets until after the evolution of its theoretical functions have given rise to science. In any case, the prediction about the perfection of the practical function of naturalistic Reason defended here uses Christianity to show how religion compensated for the superficiality of the Platonic explanation of moral goodness. But since it is not obvious what is ontologically necessary, it is less a prediction about what ontological scientists will discover than the proposal of a research project for scientists with naturalistic Reason to carry out.

The first metaphysical argument has an ontologically necessary structure because it depends on naïve realism about perception, and Athenian metaphysics constrains possible marriage partners in several ways. First, for God to define the nature of Plato's phantom Reality, God must be a substance that exists outside space and time. Though God may not have to be the creator of the natural world, only a monotheistic religion could claim to explain everything in the natural world. Second, the God of the religion wedded to Platonic metaphysics would have to be a Rational being with a free will for his purpose to be the ultimate purpose of the natural world. If God did not create the natural world because it was good, it would not pass the Socratic test of practical wisdom, and the illusion of purposiveness could not explain how knowledge of the Good as part of the True could cause believers to choose freely to do what morality requires when it is opposed by strong desires or self-interest. And third, since the natural world as perceived by direct realists is the Platonic Appearance, the Reality behind Appearance must be a supernatural Rational being whose existence reflective subjects know by Reason, and so those who believe in God would have to share God's Rational nature.[30]

The evolution of the practical offspring of the Enlightenment during the modern era may impose other constraints on possible marriage partners for Platonic metaphysics. Consider, for example, the moral equality of all reflective subjects. Since Reason is how God is known, any possible marriage partner would entail the moral equality of all Rational beings, and thus, a suitable religion may have to recognize Reason as the cognitive power that distinguishes reflective subjects from other animals, as religions in the

---

[30] This constraint would rule out Islam as an eligible marriage partner if, as some believe, it relies on force, rather than a confession of sincere faith, to win converts. The use of death threats to keep those who were raised as Muslims from renouncing their faith would also make Islam ineligible.

Judeo-Christian tradition do by holding that human beings were all created in God's image.[31] Furthermore, to explain Enlightenment belief in progress, a suitable religion may have to explain human history as a drama that comes to a conclusion, as Christianity does by its expectation of an *end of days*, rather than as endless cycle of changes, as pre-Socratics believed.

If these constraints are all necessary, Christianity fills the bill. But ontological scientists will wonder about other Christian doctrines because there is an uncanny correspondence between them and other aspects of their ontological explanation of the course of evolution. This may be just a lucky accident on Earth. But the following speculation about how those doctrines will lead Christians to recognize that God is identical to the natural world suggests concretely how naturalistic Reason will eventually serve its practical function on Earth as perfectly as it does its theoretical function.

Consider, for example, the doctrine of original sin. In the beginning, Adam and Eve were in the Garden of Eden, and though God had commanded them not to, they ate *fruit from the tree of the knowledge of good and evil* in the Garden of Eden. To Christians with naturalistic Reason, this doctrine will seem to represent war as the original sin because that was the origin of morality. War was the decisive group-level form of natural selection that enabled *modern humans*, that is, reflective subjects, to replace *archaic humans*, which were mammals using a language with only naturalistic sentences. Modern humans could see into the minds of other subjects, and since archaic humans could not, modern humans were able to outsmart them, causing the extinction of archaic humans, if only by depriving them of

---

[31] This constraint might make Judaism ineligible. Though Yahweh claimed to have created the entire natural world, his covenant with the Jews made them the "chosen people", and to explain why all reflective subjects were equal, Judaism would have to be interpreted in a certain way, if not amended. It can be argued that, in making Jews the "chosen people", God did not intend them to rule over others, like Plato's philosopher king. His commandments to Jews may have been a way of teaching others that his purpose in creating the natural world entailed a difference between good and evil with eternal significance. But since his promise was that he would protect the Jews if they adhered to all his detailed rules about how to live, the amendment to Judaism would have to explain why Jews expected the messiah that God would send to fulfill his promise to be a leader like King David, who would use state power to protect them. That makes Yahweh seem more like just one of many gods worshiped by different spiritual organisms, and it is hard to believe that such a god had created a world in which everyone is morally equal.

access to acceleration fuel. Natural selection by war gave all language-using mammals desires that motivated them to harm even animals like themselves, and thus, when they evolved reflection, the culture of their spiritual organisms had to include authoritative spiritual input that required its members to follow rules that kept its members from harming one another. Hence, reflective subjects would know the difference between good and evil, and with the ability to represent the causes of their behavior as part of the process of causing it, they would know when they chose to act on their spiritual desire and when they did not. When they chose, they would know that they could have chosen otherwise, and as beings with a free will, they would shoulder the burden of taking responsibility for choosing to resist the temptation to sin and doing what morality requires. Since Christians may see this loss of the innocence of animals without reflection as eating fruit of the tree of the knowledge of good and evil, they may see the original sin as war.

As we have seen, the illusion of purposiveness is inherent in reflection, and the Christian gospel can persuade reflective subjects that the meaning of life has to do with choosing to be moral even when it is not easy. Christians were required to love their neighbors as themselves—even when they were enemies. The Christian gospel is a tantalizing, rich, and inspiring story about a perfect moral hero. Christ taught his followers to love a God who forgave their sins, and when his neighbors crucified him for teaching this, he even forgave them for what they did. His resurrection from death promised the ultimate vindication of his moral lesson, and the promise in *Revelations* of a second coming of Christ was the belief that the ultimate outcome of this drama was heaven on Earth. This eschatology was a distinctive trait of Western religion.

This may not be necessary for the Enlightenment belief in progress to evolve. But since evolution by the reproductive mechanism is progressive, the explanation of Western civilization as the metaphysical stage of evolution implies that its history is progressive, and it is uncanny how it solves the most profound dilemma of Christian theology. Christianity claims to be a monotheistic religion, but it holds that God was three persons. The doctrine of the trinity seems blatantly contradictory, and it long invited mockery because since theologians have never found a satisfactory way to explain it. But with its discovery of the identity of God and nature, naturalistic Reason can remove the apparent contradiction.

The Christian God was supposed to be three persons: the Father, the Son, and the Holy Spirit. Though theologians claimed that they were all constituted by a single substance, this pretense did not pass the Pre-Socratic test of ontological coherence. The Father and the Son differed from one another as a God outside space and time and a God who is an object in space, while the Holy Spirit was just a magical power that enabled the Father to reach across the dualism of Being and Becoming and act in the natural world. But if God is identical to the natural world, the doctrine of the trinity has a coherent ontological explanation. If the substances constituting the natural world are God, God as the Father is the first cause because those substances cause the existence of everything in the natural world. If the reflective subjects who acquire naturalistic Reason are the Son, God as the Son is a Rational being who knows the True and the Good and can die on a cross. Finally, if naturalistic Reason is the Holy Spirit, God as Holy Spirit is how God acts in the natural world because it is the neural BGS by which the spiritual organism guides the behavior of its members after the Second Enlightenment. Thus, if the Christian God is identical to the natural world, the Father, the Son, and the Holy Spirit are all constituted by the same substances.

If the metaphysical stage is another inevitable stage in the evolution of life, evolution leads inevitably to reflective subjects in whom Reason serves its theoretical and practical functions perfectly. Thus, two persons of the Trinity come into existence as time passes. But that does not mean that the Christian God comes into existence in the natural world. All three persons of the Trinity always exist. God as the first cause is spatially infinite, and since only the present exists, God as the Father always exists everywhere. Since God is constituted by space and matter, planets suitable for the evolution of life exist throughout an infinite universe with evolution beginning on them at different times. Thus, every essential development in the course of the evolution of reflective subjects with naturalistic Reason exists somewhere in the universe at every moment in the passage of time. So, the same substances constitute all three persons at every moment in the passage of time.

Furthermore, reflective subjects who acquire naturalistic Reason will notice that evolution gives the natural world traits that explain all the perfections that theists traditionally ascribed to a transcendent God. God is *omnipresent* if God is identical to the natural world because space and matter constitute everything in the natural world and they endure through time. God is *omniscient* if God is identical to the natural world because after the Second Enlightenment, reflective subjects will know everything that holds necessarily in the natural world, and though they will not know everything that happens, they will know the causes that explain it. And God is *omnipotent* if God is identical to the natural world because when naturalistic

Reason serves its theoretical function perfectly, spiritual organisms will be as powerful as any organism in the natural world can be. Finally, God is *absolutely good* if God is identical to the natural world because when naturalistic Reason serves its practical function perfectly, spiritual organisms will choose goals that are good not only because they are the goals that its BGS normally chooses but also because they contribute to the natural perfection of life. Indeed, since theists believe that God gave human beings dominion of over nature, *absolute goodness* can be defined as what spiritual organisms choose when reflective subjects have naturalistic Reason and it serves its practical function perfectly.

As we have seen, for all other BGSs, goals are good because they are the goals that their BGS normally chooses. Since evolution by the reproductive mechanism is progressive, the goals that are good for a BGS are increasingly powerful in controlling conditions that maximize its functional power. That is how geometrical action aimed at the good contributes to the natural perfection of all the forms of life in any region on the surface of a planet. (In any region, the limit approached asymptotically by progressive evolution is using as much of the acceleration fuel available there as possible to fuel geometrical action aimed at the good.) But these effects of BGSs are not why goals are good because contrary to the illusion of purposiveness, goodness is not a cause of behavior. Rather, goodness is an essential aspect of life, because living organisms are choosing machines and choosing goals that are good is what matters to choosing machines. To be sure, goodness is relative to specific forms of life. But naturalistic Reason is a unique kind of BGS because it knows the nature of life, it knows that goals are good because they are chosen by BGSs serving their function normally, and it knows that it is the BGS whose evolution during the metaphysical stage changes in the direction of maximum power to control conditions that make it possible for spiritual organisms to reproduce if and when the time comes. Using this knowledge of its own nature as a BGS to serve its behavior guiding functions is what makes naturalistic Reason perfect at serving its functions, both theoretical and practical. Since it knows the direction of evolution caused by the reproductive mechanism, it can know which of the goals that it might choose would contribute most to natural perfection. Uniquely in the case of naturalistic Reason, therefore, the goals that it normally chooses will coincide with the goals that contribute most to natural perfection. The goodness of its goals will not be relative to a particular form of life but, rather, to a form of life that knows the True and the Good. Naturalistic Reason will be absolutely good.

Indeed, its choices will depend on knowing the Beautiful as well as the True and the Good because reflective subjects with naturalistic Reason will take up where the reproductive mechanism leaves off. Part of choosing the goal in any situation is choosing the best means to ends, and since the principle of optimality identifies the best means to goals that are good, naturalistic Reason can use that principle to identify the Beautiful as optimal means to ends when the end does not depend on its being good. The ideal is still doing the most with the least in the way of causes. There is a fact of the matter about optimal means-ends relations, and Rational beings can know it by comparing alternatives in their faculty of imagination. Thus, using the principle of optimality to define the nature of what is beautiful in that way, naturalistic Reason will choose goals that contribute even more to change in the direction of progressive evolution than can evolve by natural selection. Choosing goals that are beautiful is artificial selection. But it is also cultural evolution by aesthetic selection, and when Rational beings use the Beautiful as part of the Good and the True to take up where the reproductive mechanism leaves off, naturalistic Reason will determine the future course of change on its planet

This picture of where the metaphysical stage is headed is heady stuff. But the end is a long way off. The Second Enlightenment begins with the discovery of how God is identical to the natural world, and though ontological scientists will know the direction of change by which it is completed, even they will not know what happens after the metaphysical stage. That will depend on imagination and judgments about what is most beautiful. All that a science based on ontology can foresee is the kind of perfection that evolves during the metaphysical stage. On Earth, it was foreseen through a glass darkly as the perfection of a transcendent God, and when ontological scientists discover that God is identical to the natural world, they will see the completion of the Second Enlightenment as Earth embodying all the perfections of God. Though some aspects of the completion of the Second Enlightenment are predictable, naturalistic Reason cannot predict what God will choose to create. That is a mark of creativity.

Nor is there any obvious end to God's work of art. Spiritual organisms may be powerful enough to retreat to a colder planet when the Sun becomes a red giant about 4 billion years from now and destroys Earth. And when the red giant becomes white dwarf star, they may be powerful enough to come back in close and use it as acceleration fuel for life to exist endlessly into the future—if that is beautiful and makes the most out of ontological causes.

Conclusion

Naturalistic Reason presents the reasons for believing that the discovery in physics about space being a substance that interacts with matter will cause a revolution in which beings like us begin to see the evolution of culture by Rational selection as bringing about heaven on Earth. *Sapere Aude* dares readers to consider that argument. And it *will* take courage because what it predicts seems too good to be true. Indeed, in the current era, it defies commonsense. Everyone knows that there is no such thing as the True, the Good, and the Beautiful. That was just foolish hokum passed down to use from a bunch of dead white men in Western civilization. But *all* it will take to make this prediction true is courage. If you have come this far, you are able to follow arguments, and all *you* have to do is consider an argument that claims to show that naturalistic Reason can know Reality behind Appearance.

Sapere Aude is just a summary of an argument that shows, in the near term, how all our basic disagreements will all be resolved. It begins with science, whose way of knowing almost everyone considers reliable, and it uses the most basic conclusions of science to show how a series of plausible discoveries will lead to a scientific explanation of every kind of thing found in the natural world. Eventually, everyone will recognize that they exist as part of a perfect being. But *Sapere Aude* is just a summary of the argument, and such a high-altitude view leaves plenty of room for doubt. Many possible doubts are addressed in this trilogy, as indicated in *Part Three* by the tables of contents for its three volumes. And there are two pathways through the trilogy: a short one in which mountain tops are linked by high-speed bridges, and a longer one in which paths lead down through valleys to the next mountain top. The longer path turns up some problems about details that I am not sure how to solve, and I leave them as research projects for others to carry out. But it is hard to see how the discoveries that solve these problems could cast doubt on the basic discovery about the natural world being constituted by space and matter as substances interacting with one another as they endure though time. This book sums up the predictions based on that ontological theory, and if they are contradicted by what perception finds, this argument will be easily and quickly refuted. But in the absence of obvious objections, we can expect to find solutions to the problems encountered, so it is reasonable to be optimistic about the beginning of the cultural change predicted here.

In the absence of the courage to consider this argument, on the other hand, the reasons for pessimism about what will happen are dreadfully compelling. The American spiritual organism is dividing into opposing tribes. Its culture is regressing to the psychological-level stage of spiritual evolution without the help of a shared religion for choosing which goals to pursue jointly. Political tribes disagree not only about values but facts as well, including even facts about current events. And without a way of talking to one another about what is happening that can resolve our disagreements peacefully, the only possible resolution will be deception and violence. Things are falling apart. The center may not hold. And even in America, might is becoming what makes right.

As it stands, no one else claims to have an argument that promises to settle all our basic disagreements, and this is basically a simple argument. It predicts that a single discovery that physics can make now will begin a revolution in which Reason eventually prevails. If the reasons for the predictions are sound, they plot the course of this cultural revolution in considerable detail. But the labor of carrying it out will fall on the shoulders of others. Discoveries are predicted in every intellectual field, and experts in each field will naturally look for ways to show that the puzzles confronting them are not solved so easily. I believe that they will fail to find decisive objections to these predictions. But in their search for them, I believe that they will find their conflicts with experts in other fields vanishing. Experts will understand more clearly how their fields depend on one another, and instead of devising special methods to use in their fields, they will find themselves drawn into agreement on basic issues. In the absence of empirical falsification, they will be unable to resist the coherence of naturalistic Reason. This revolution in culture will not be political. It will not ask anyone to submit to any authority but their own independent judgment about what is true. To be sure, those who do not care enough about the true to argue about it will, as always, accept what others around them believe, and like children, they will simply go along with those who do take arguments seriously. But everyone will come to have beliefs that can be shown to be true by their correspondence to Reality, and since they will agree about the nature of Reality, everyone will agree with one another.

The 18th-Century Enlightenment arouses nostalgia about confidence in Reason, and though the number who even know about the Enlightenment is dwindling, this argument offers hope to everyone who still believes in the rational pursuit of truth. It shows how science, which is our most reliable kind of knowledge, can become naturalistic Reason knowing Reality behind Appearance. It will take time and effort to consider this argument. And given the disillusionment with Reason in science, it will also take courage. But it will not be reasonable to refuse to muster courage on the grounds that it is foolish to take such an unlikely argument seriously. Even to disagree with one another, we must share a world, and it is hard to believe that it does not have a single consistent nature at bottom. And the need for a *Big Picture that Everyone Can Share* is too great not to consider the possibility that one is available. In the near term, your courage will be paid off by pushing endarkenment back into the sea of passionate relativism from which it rises. And in the long term, the payoff will be so great that it is not even conceivable now.

Part Three

Tables of Contents of the Trilogy

Part Three uses the *Tables of Contents* of the three volumes of the trilogy to show where further discussions of specific issues can be found.

Naturalistic Reason is formally the prediction of an unexpected revolution in culture. It gives reasons for believing that it will happen, and they are detailed enough that, if they are sound, they might even bring out the revolution they predict. But the truth of its assumption and the validity of its logic is something that everyone must judge for themselves, and those who find the argument plausible are still likely to have doubts about some steps along the way. Since *Sapere Aude* is itself a high-altitude summary of the argument, there is no index to it. But more detailed arguments addressing every issue can be found in the three volumes summarized here, and their tables of contents, reproduced in Part Three, can be used to locate arguments addressing specific doubts.

Though reading all three volumes takes considerable time and effort, the burden is not as great as it may seem at first because there is an abbreviated pathway through each of them for readers who are not concerned with the technical issues on which professionals in the relevant fields will focus. It is much shorter, especially in the first two volumes on physics and specialized sciences. It is like a highway for tourists with bridges that lead from one mountain to the next, and for those who need to go into the weeds, there are junctions (marked by signposts, ⚙) where less traveled roads lead down through the valleys to the next mountain and eventually rejoin the highway.

No one can be confident about this argument unless specialists in relevant fields are convinced by what they find—or are able to discover for themselves—in the valleys. But readers following the shorter pathways will see where technical issues arise. In *Volume I*, the approximate length of the abbreviated pathway is 215 pages (86K words), while the complete pathway for more technical readers is 490 pages (196K words). In *Volume II*, the shorter pathway is about 174 pages (70K words), while the longer path is 498 pages (200K) words. And in *Volume III*, the pathway for general readers is 380 pages (150K words), while the pathway for technical readers, at 405 pages (165K words), is not much longer. They all have indexes and relevant references to other publications.

Finally, there is also a website, *natReason.com*, that provides information about resources for those who are interested in considering naturalistic Reason. Along with links to sites where the books can be purchased, there are links to a series of videos that also give an initial view of the overall argument. Videos addressing specific issues about what naturalistic Reason implies will be added as the need arises.

Chapter 1

Unification of Physics (Volume I)

Table of Contents

Chapter 2

Unification of Science (Volume II)

Table of Contents

Chapter 3

Unification of Science and Philosophy (Volume III)

TABLE OF CONTENTS

About the Author

Phillip Scribner was born in Denver, Colorado and reared mostly in Cheyenne, Wyoming. He was graduated *magna cum laude* from the University of Colorado in Boulder in 1961 with a major in philosophy, and he received his Ph.D. in philosophy from Johns Hopkins University (Baltimore, Maryland) in 1966. After teaching at the University of Nebraska in Lincoln for four years, he taught philosophy at the American University in Washington, D.C. for thirty years. He started writing the argument that has become this trilogy around 1975, continued working on it after retiring from teaching in 2000, and completed it in 2022 at the age of 82.

Made in United States
Troutdale, OR
11/03/2026

56844682R00091